The Human Nature of the Singing Voice

Born in 1946, Peter T Harrison began his musical training as a chorister at King's College Cambridge under the direction of David Willcocks. He went on to study singing with Joy Mammen at the Royal Academy of Music, London, where later he taught in her place for two years. He was a professor at the Guildhall School of Music and Drama, and has given lessons and workshops in many parts of the world. He has also conducted classes for singing teachers. In 1994, at the Third International Congress of Voice Teachers, Peter was selected as one of seven 'Master Teachers'. He has been a member of the British Voice Association since its beginnings in 1986.

Since 2000 Peter Harrison has been Resident Professor and Director of Vocal Studies for the Estúdio de Ópera do Porto, Casa da Música, in Oporto, Portugal.

THE HUMAN NATURE OF THE SINGING VOICE

Exploring a Holistic Basis for Sound Teaching and Learning

Peter T Harrison

Dunedin Academic Press
EDINBURGH

Published by
Dunedin Academic Press Ltd
Hudson House
8 Albany Street
Edinburgh EH1 3QB
Scotland

ISBN 10: 1 903765 54 4
ISBN 13: 978-1-903765-54-8

British Library Cataloguing in Publication Data
A catalogue record for this book is available from the British Library

Typeset by Makar Publishing Production
Printed in Great Britain by Cromwell Press

To my children Giles and Claire

To the memory of John Riley-Schofield (1954–2005)
a pupil and friend who shared my passion for great singing

To all those still wondering

Contents

Preface xi

Acknowledgements xii

Illustration Credits xiv

PART I – **Civilisation of the Human Voice**

CHAPTER 1: ***Sounds of Life*** 3
Introduction – The primacy of the ear – Aural perception and the voice – Musical, tonal and aural imagination – Hearing versus physical or visual information – The natural vocal monitor

CHAPTER 2: ***Sounds Intelligible*** 11
Intelligent strangers – The need to voice our feelings – Hijacked voice – Voicing text – Faulty speech transmission – The mind-body split – Laryngeal collapse and support – The emotional connection – *Passaggi* – Only for singing – *Messa di voce*

CHAPTER 3: ***The Stifled Cry*** 23
The repression of emotions – Self-revelation and the power of singing – Emotional and vocal physiology – Demons, diamonds and guilt

CHAPTER 4: ***Bowing to Life*** 29
Posture and how the structure of the body suffers from neglect and misuse – Self-determination and the human spirit – Reasons for not giving the body due credit – Adverse effects of a sedentary life-style

CHAPTER 5: ***The Inspiration of Life*** 35
The relationship between the larynx and breathing – The experience of breathing out and breathing in – The relationship of breathing and singing – Breath control and management – The importance to freedom of having a sense of scale

CHAPTER 6: ***Unavoidable Conclusions*** 46
Individual capacity – Vocal liberation and personal exposure – The natural voice debate – Compensating for incapacity

PART II – **Sounding the Self**

Chapter 7: ***Making Connections*** 55
Connecting uncoordinated spheres of vocal action – The importance of antagonistic forces – Defining the singing voice in terms of its structure – Redefining 'support' and *legato*

Chapter 8: ***Mixed Objectives*** 61
A unique instrument requiring unique treatment – The crucial difference between vocal training and making music – How the two activities can complement or confuse one another – The process of *un*-mixing objectives

Chapter 9: ***Vocal Liberation*** 67
Towards a definition of vocal liberation – Healthy design and training – An individual journey – Dedication and radical treatment – Tension audible and invisible – Vocal athletics

Chapter 10: ***Training Ground*** 74
Larynx – Suspensory mechanism – Breathing apparatus – Diaphragm – Posture

Chapter 11: ***Hearing Our Way*** 88
Outlines of an 'aural picture' – Mixed voice – The missing link – Beginning the aural quest – Neutral throat as prerequisite – *Portamento* – Breath and breathing – Energy, vibration and movement – Sounding emotion – Vowels and 'placing' – Mimicry

Chapter 12: ***Trial and Error*** 103
The need for process – Creative tension – The spiral of work – Imagination and perception – Tone sense – Imagery and hearing – Hearing and physical sensation – The value of play – Voice first – Laughter – Gestures, movements and inner rhythm – Embodiment of feelings – Progressing towards music – Improvisation of means – Hearing versus thinking – Attention, distraction and confusion – Sequences and starting points – Types of vocal sessions – Various Italian terms

Chapter 13: ***Muscle Training and Fitness*** 129
Working with muscles – How to approach training exercises

Chapter 14: ***Singers' Health*** 136
General health – Ingestion – Digestion – Weight – Water – Dryness and the mucous membrane – Immunity – Breathing – Sleep – Stress and worry – Talking

Chapter 15: ***Gurus or Guides? (Teaching)*** 142
Introduction – Responsibility – The teacher's voice – Talent and trainability – Relationship and trust – Communication – Healthy criticism and evaluation – The voice rings true – Specialist knowledge – Fresh beginnings – The 'feel good' factor – Barriers to progress

Chapter 16: ***A Clean Slate (Learning)*** 156
A personal journey – Equipment – Technique – Scale and skills – Tackling contemporary music – In between lessons – Living rhythm – Hearing and other sensations – Seeking models of healthy vocalisation – Six obstacles to liberating the singing voice – Posture revisited – Assessing the teacher

PART III – **The Communicating Imperative**

Chapter 17: ***Attributes of the Liberated Voice*** 177
Introduction – The voice's innate skill – Singing joyfully? – Some of the qualities and characteristics of a liberated voice

Chapter 18: ***Attraction and Repulsion*** 187
The voice and communication – Projection – Sexuality and communication – The listener – Emotional involvement

Chapter 19: ***Being Fully Prepared*** 194
Stage fright – Learning to perform without singing – Criticism – Coping alone – Spontaneity – Learning, memorisation and preparation

Chapter 20: ***Going Deeper*** 207
The source of the singing voice – Before words – Musical authenticity and the singing voice – The diminishing power of singing – The ethics of performance

Chapter 21: ***The Ego and the Egoist*** 213
Self-esteem, sharing and creativity – The need for prima donnas – Training for survival – Pigeon-holes and tessitura – Repertoire – Specialisation and the authentic self

PART IV – **Back and Beyond – Redefining *Bel Canto***

Chapter 22: ***Healthy Communication*** 221
Being naturally expressive – Heart and mind in sound – Being fully alive – The simplicity of being – The vital present

Chapter 23: ***Redefining* Bel Canto** 225
What happened to *bel canto*? – *Bel canto* as a basis for progress – Benefits of psychology – Self and other – Collective unconscious – Gender balancing – The idea of perfection

Chapter 24: ***'Beauty is Truth, Truth Beauty'*** 231
Equating beauty with truth – Looking for the beautiful – Being the truth – Taking care of beauty and truth

Postlude 234

Endnotes 236

Bibliography 241

Preface

In this book I explore the singing voice and its relation to life, art and communication. Writing from a teacher's point of view for anyone who deals directly or indirectly with singers, I attempt to answer the critical question 'what are we actually dealing with?' In answering this question our teaching and care can assume a logical process which recognises and makes the most of individual potential and quality while avoiding common pitfalls, practices and diversions which are contrary to our voice's nature.

My observations over 35 years of teaching have taught me that understanding and working with the human voice has as much to do with its natural context as with its artistic employment. This has led me to the holistic view of voice and singing which I express here in both practical and philosophical terms. My hope is that this approach, understood as a whole, will provide a stimulating basis for those who require a different perspective from the usual strictly scientific or academic one for the development of their work in this fascinating and life-enhancing field. Above everything else I appeal to the evidence of the human ear and the spirit of wonder in the soul-searching quest which is teaching and learning singing.

Acknowledgements

It was my good fortune to study singing with three outstanding voice teachers: Yvonne Rodd-Marling, who worked alongside Frederick Husler and was instrumental in producing their still-popular classic *Singing: The Physical Nature of the Vocal Organ*, and Rodd-Marling and Husler's pupils Joy Mammen and Pieter van der Stölk. To these wise, caring and skilful teachers I owe an incalculable debt of gratitude. Those who understand the work of Husler and Rodd-Marling speak not of a technique but of a 'school'. It is the liberating philosophy behind the Husler-Rodd-Marling approach to the human voice that has been the inspiration for my own work and development for almost 40 years.

I have learnt most about teaching from my pupils, who have not only provided me with a fascinating and rewarding livelihood, but have presented me with the intriguing challenges which collectively have led to the writing of this book. There are too many to thank here individually, but I do thank all of them wholeheartedly, from those who have tried me with their seeming unwillingness or inability to respond to those whose voices have begun to flower under my guidance and made the effort so worthwhile.

Thanks are well overdue to all those who have been enthusiastically influential in the shaping of my teaching career, among them David Willcocks, Joy Mammen, Noelle Barker, Susana Herner, Monica Preux, David Wilson-Johnson, Delia Lindon, Scilla Askew and Luis Madureira, and to those friends who have most nourished my thinking: Nicholas Bradbury, Richard Hames, Maria Fernandez, Lorna Marshall, George Hadjinikos, Gillian Patterson, Piluca Tejero and Ana Guiomar Macedo.

Thank you to all those who have lovingly encouraged and supported me in the writing of this book. I list some of the most avid! James Dooley, Pilar López, Sara Stowe, Teresa Villuendas, Mary Wiegold, Brian Gordon, Annabella Waite, Lucia Lémos, Janet and Laurie, Luís and Carmen, Kathy and Marcellus, Cecília and José, Elise and Joost, José and Paula, Graham and Liz and all those who have most closely shared my life.

Thanks are also due to my publisher's wife Veronica who saw my original proposal for this book in Dunedin Academic Press's rejection pile and persuaded them that this was a book well worth publishing!

Finally, a huge thank you to Kate Hopkins, my editor from the first draft to the last, for her unfailing enthusiasm and encouragement, especially throughout the tricky and often painful process of cutting my original manuscript down to size. It is difficult to imagine how this book might have materialised in any satisfactory shape or form without her constant attention and feedback.

Although my ideas for this book derive mainly from my own teaching experience and observations, I am sure that many have already been expressed in some way or developed to some degree elsewhere. I therefore apologise in advance if I have inadvertently failed to attribute or have misattributed any of the ideas expressed or material reproduced. Anyone who has reason for concern in this respect may contact the publishers so that omissions or mistakes can be righted in any future editions.

Illustration Credits

The four part-title drawings are by Piluca Tejero.

Illustrations 10.3–10.9 and 10.12 are adapted from *Singing: The Physical Nature of the Vocal Organ* by Frederick Husler and Yvonne Rodd-Marling (originally London: Faber & Faber Ltd 1965) revised edition London: Hutchinson 1976, pp. 18–34, by kind permission of Tremayne Rodd. 10.3 is originally by K Goertler. 10.5–10.9 are adapted by Alexa Rutherford.

All other illustrations are by Ruth E Scott and Peter D Scott.

PART I

Civilisation of the Human Voice

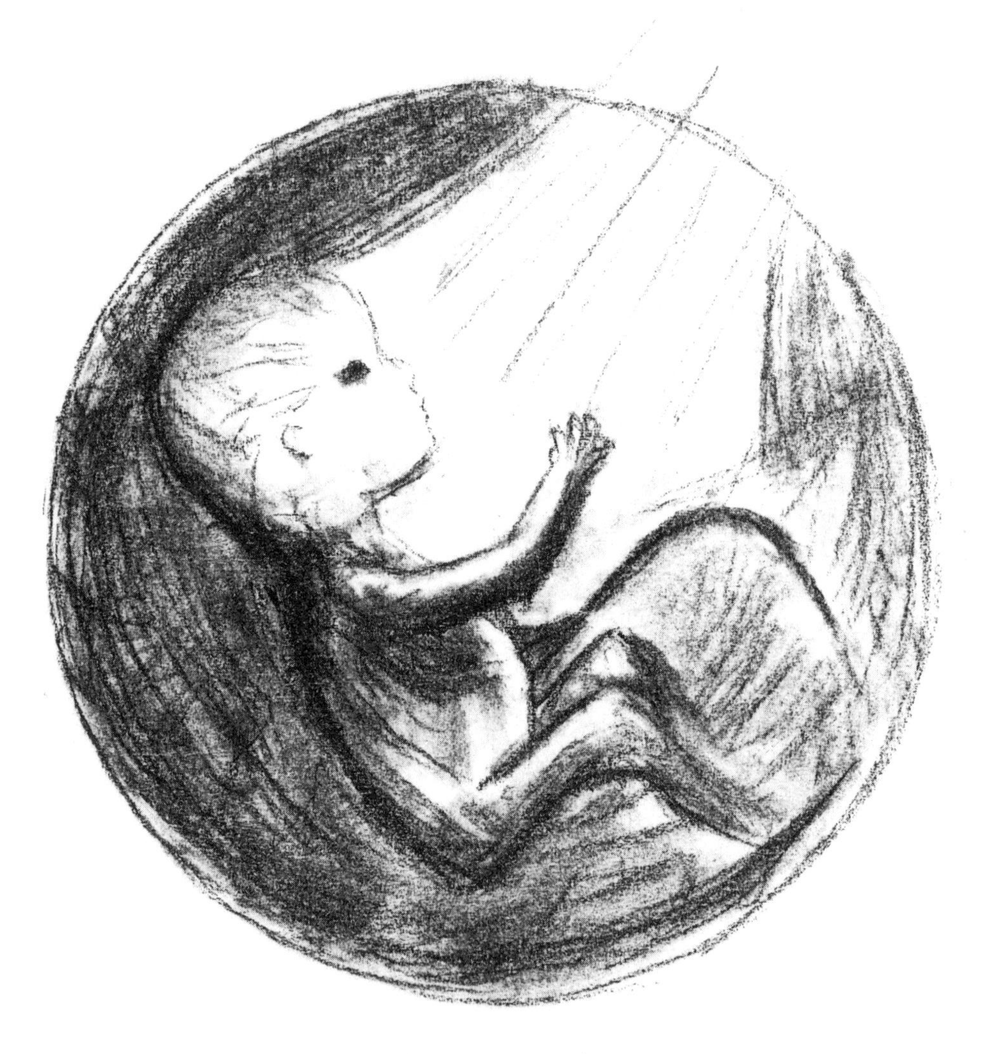

Human Potential

PART I

CHAPTER I

Sounds of Life

Introduction

Before any of us uttered a word, before we drew our first breath, we developed a faculty more important than speech for without it we would have remained mute. Phylogenetically ahead of all our other senses (developing in the 20th week of foetal life), hearing takes pride of place in our evolution. Why? I believe it was primarily our hearing that guided and informed our evolution, ensuring our survival as the human species.

In *Hearing Equals Behaviour* Guy Bérard states 'Everything happens as if human behaviour were largely conditioned by the manner in which one hears'.[1.1] Bérard demonstrates that conditions thought of as psychological, such as learning disabilities and depression, can be caused by aural impediments. Could it be also that *what* we hear and *how* we hear it are significant determinants of good health over and above the norm? Could hearing ourselves, and the *way we hear* ourselves and the 'self' in others be significant psychologically and sociologically?

In the comfort of our mother's womb, we bask in a watery sound world which provides our first environmental impressions. Even at this stage, we are selective in our musical taste – a five-month old foetus responds to sound and melody in discriminating ways.[1.2] Could the sounds we hear *in utero*, and the way we receive them, have a bearing on our subsequent emotional life? Sound affects us before anything else. It provides our first palpable contact with the world. Once born, we quickly recognise and readily respond to our mother's voice, and soon 'discover' our own. Through aural interaction we learn to distinguish one from the other, thus gaining a first unthreatening sense of separateness. When Mother is not physically close, we call and she responds – she is within *earshot* and all is well.

Clearly, the auditory system profoundly affects our relationship with the outside world. Sounds arouse our curiosity, we hear in the dark and from a great distance, and our instinctive fears, of loud noise (alarm) and falling (loss of balance) are bound up with our aural system. We absorb sounds unconsciously, and rapidly build up a vast sound-vocabulary.

As a baby it seems we come across our voice by chance, sometimes feeling entirely at home with it, while at others being surprised by it. It becomes gratifying, and we begin to exploit its power. When, as a child, we hear a sound, we mimic it unhesitatingly and effortlessly, and at first we make no distinction between speaking and our more lyrical vocal excursions; it's all sounds to us.

—

In spite of the importance of our hearing, and the fact that singing is essentially sound, the idea that to be in control of one's voice one must actually *hear* it, is perhaps the least understood aspect of singing. Singers are generally asked: 'What does it *feel* like when you sing?' rather than 'What does it *sound* like?' They are asked what they *do* when they sing and a physical description is expected.

A singer does have a physical sense of what he's doing, just as a builder or a dancer does. However, both teachers of singing and singers might profitably ask:

1. Of what benefit is the physical sensation to the singer?
2. Is this sensation indicative of good practice?
3. Do all singers feel their singing in the same way?
4. Can physical sensations be depended upon?

However strongly felt, physical sensations are invariably subjective. How do singers know that what they are feeling is what their teacher wants them to feel? How does the teacher know that what the singer is feeling when she sings is the same as he feels in his demonstration? Looking for specific sensations when singing is certainly a popular pursuit, but it is also the subject of much disagreement.

To control our voice by means of our hearing may seem a nebulous alternative. One of my colleagues once asked: 'How do you know that what you are hearing and what I am hearing are the same?' My answer was, 'I don't'. However, while we may have been listening *for* and *hearing* different things, there is no doubt that we were listening *to* the same sound.

Some teachers find the idea of listening so fraught with difficulties that they maintain that a singer should not use his ears at all! How then should a singer 'hear'? There seem to be four main areas of concern:

1. A genuine inability to hear one's own voice

The inability to hear and thereby to monitor one's own voice is a kind of deafness, a disassociation of sound from its source. The innate neurological communications system linking the ear with the larynx has broken down or become confused. We rarely notice this until we begin to think or worry about singing, and even then we rarely appreciate the extent of the problem or how it arose. As will be explained later, the ear has lost its primary singing orientation and instead got used to the superimposed patterns of speech. It has forgotten how to sing!

Endeavouring to find the singing voice when the ear is so attached to deeply ingrained speech patterns can make it seem remote or 'foreign'. The saddest outcome is that singing is experienced and seen as something unusual, for which one must be specially gifted. In medical and scientific circles the singing voice is commonly thought to be an extension of speech rather than something pre-dating it which has deteriorated in speech's favour.

Our 'vocal deafness' is exacerbated by the fact that it is with speech that we have gained our vocal identity, our identity in sound. This is why at first we may not recognise ourselves in our singing voice. If we want to regain this voice, we have no alternative but to reacquaint ourselves with it by hearing the differences between our speaking and our singing voices and allowing ourselves to recognise our 'being' in a relatively strange sound.

2. The need for security

The view that singers are insecure is widely held. Perhaps they are insecure with good reason – their voices are frequently abused and treated like machines, to be switched on or off at will. Singers are often in fear of their health or derided for being unmusical. I once heard it said that an eminent conductor thought that singers should be treated like lemons and squeezed to the last drop! What singers do and how they do it is rarely understood by other musicians, critics or audiences. The feeling of insecurity is uncomfortable and a strong motivator; singers understandably want to feel 'in control'.

Any performer who is ill-prepared or has an unreliable memory may feel insecure. An ill-prepared performance can provoke damning criticism but, however well-prepared we are, we never actually know what's going to happen from one moment to the next. Music is not mechanical but needs room to breathe and freedom to flow. Trying to pin it down is like trying to capture running water in a bucket, losing the very quality by which it is best defined. The more civilised we become the more systems of security we devise, but life is inherently insecure so perhaps we would do better to learn to ride the wave like a surfer who studies and respects nature and, alert and vulnerable, strives to become one with the wave. A good singer gives in to the inevitable fluidity of musical phrases and rides them with confidence.

The physical sensations we seek when singing serve as props, something to grasp so that we feel something concrete we can depend upon. We may then feel that we can 'produce the goods' without faltering. These props, however, take the form of tensions and exaggerated movements that limit our voice's flexibility and range of musical and emotional expression (thus rendering spontaneous music-making practically impossible). They tend to become more imperative and less effective as the years, sometimes not many, go by or as the weight of professional work increases. Eventually, either the voice breaks down, or another 'more helpful' sensation or prop is sought. Such measures are usually referred to as 'techniques'. Designed to enable the singer to perform in spite of the vocal problems or inadequacies that might otherwise undermine his confidence, they are invariably deliberate physical manoeuvres. Techniques are, at best, a bit of a lottery. The only person who really knows if a technique 'works' is the one who's gaining satisfaction from it. To have the same feeling experience, we would have to share the same body.

It is different with the singing sound. Whatever sound you make, in whatever way you make it, is the same sound that I hear. Anyone may measure a singer's vocal performance against what he, subjectively, considers to be good singing or, more objectively, against what the requirements of the music appear to be, but true judgements about the efficacy of a singer's technique can only be made through what is heard – the ingredients of the sound. This is not to deny the vibrations transmitted by the singer and sympathetically felt by his audience. Whatever the result in effect or intelligibility, deliberate physical work, felt by the singer in specific locations of his body as part of a practised technique, whether evident in exaggerated facial distortions or other extraneous gestures, is distracting, and can cause the most sympathetic audience a good deal of discomfort. The would-be singer's challenge is to reach a point at which she can dispense with conscious techniques and concentrate instead on what she is communicating.

3. The fact that the voice and its processes are largely invisible

The 'correctness' of a voice is often judged by what can be seen, never mind what cannot. The well balanced voice, however, comprises interacting muscle movements that reach down to the pelvic floor and as far up into the head as muscles can go. During the course of a phrase of music this widespread system remains intact and in movement from beginning to end. Although almost imperceptible, this total movement is clearly audible. It may be the invisibility of the muscle movements that produce the sung tone that drives teachers who cannot trust their ears to want to *make visible*, as well as felt, what is going on. Singing teachers can fall foul of feelings of insecurity too! The 'need' for visual proof has been a bane of voice research laboratories since Manuel Garcia Junior invented the laryngoscope.[1.3] Thus far no equipment designed to measure the voice possesses the comprehensive diagnostic skills of the human ear.

4. Confusion about *what* one's supposed to hear, or for what one should listen

There are various possibilities:

The sound that returns to us from the space in which we are singing?
In a resonant room the echo can be seductive or distracting. Giving in t temptation to listen to oneself from the outside (including cupping a hand fr mouth to ear) causes hesitation from delayed feedback and consideration of the result. Those put off by this delay may stop listening altogether, preferring to rely on what can be experienced physically. A dry acoustic, however, though neither seductive nor distracting, may be disconcerting, making a singer try to *produce* the resonance which seems to be lacking.

When as a student I sang a solo at St Paul's Cathedral it seemed as though I had lost my voice, or that it had flown straight into the great dome. There was no appreciable feedback. Later I was assured that I'd done a good job, but this startling lesson taught me that acoustics are not to be relied upon.

Vocal tone is irretrievable and continuously renews itself in response to our imagination, the sentiment we are expressing and the dictates of our musical mind. The more freely it is emitted, the more speedily it flies – the sung tone will not, under any circumstances be pinned down. Conscious attempts to control it invariably lead to some kind of physical inhibition or struggle, with attendant limitations.

In listening for external feedback we become mentally preoccupied with the sound we are making or rather have just made, attempting to adjust it when it has already passed, and thereby further impeding vocal fluency. In singing, our mind's rightful place is ahead of our voice, concerned only with what inspires or informs the performance and trusting that our voice will follow suit.

Should we be making sure that we are audible among other instruments and voices?
Successful music-making demands a high degree of aural clarity, sensitivity, and discrimination with regard to what's going on around the musician. However, as many singers find to their cost, the sound surrounding them can seem to blot out their own. This can be alarming and disarming. If the singer loses her cool in these circumstances, her insecurity may urge her into competition with other singers, piano or orchestra.

Resorting to forceful measures (that in a less overwhelming situation a singer might not dream of employing) is likely to prove counter-productive. The singer is inclined to seek physical 'help': extra effort, felt to be necessary and appropriate in the circumstances. This is a long way from the sure-footedness that can be gained from the re-establishment of those ear to voice communication lines, when the singer no longer has to concern herself with the 'how', but is free and sufficiently confident to get on with the 'what' of her performance .

Electronic feedback, even if it's more direct for the singer than an echo and succeeds in segregating her voice from whatever else is going on, is not the real, raw sound being made. It's filtered and at least one remove from the singer, who not only gets a false idea of balance but once again hesitates in considering the sound. Singers are often surprised when they hear their recorded voice. Sometimes this is sheer lack of recognition, but, more importantly it isn't what they heard *while singing*, and will not be what they hear when they sing again.

Having more or less dismissed both the hearing of external feedback and physical sensation as unreliable criteria by which we can best know and guide our voice, we now move another step closer to the kind of aural awareness which, although it may be hard won, is foolproof in its design.

Without **musical imagination** our singing would never get off the ground. Although in some it may have gone to sleep, musical imagination seems to be indivisible from the desire to sing itself. It may need stimulating or feeding but it is there in readiness; witness the vocal explorations of babies and young children and the instinctive phrasing of so-called 'natural' singers. If we want to make a good job of singing, our musical imagination must be up and running. Someone described as having a 'musical ear' clearly manifests this faculty: he has in his mind's ear a clear vision of what is about to be played or sung, and, running ahead of his playing, his imagination not only keeps the sound spinning but facilitates creativity. This anticipatory quality of our imagination – possibly even more important for singers than for other musicians – must be assiduously cultivated, since in true singing there is no place for artifice. Contemplating how the voice is being produced is akin to an instrumental musician adjusting the structure of his instrument as he plays; the flow of tone (and therefore the musical line and continuity of intention) is in constant jeopardy.

Tonal imagination (being able to anticipate colour and quality of tone) is similar. It too needs to be highly developed, especially in view of the organic nature of the singer's instrument. The tonal palette, like the basic quality of voice, is unique to the individual singer, and becoming acquainted with it leads to aural *self*-awareness. Tone sensibility may also be linked to our desire to sing. Inspired by what they want to 'say' to small offspring, adults spontaneously combine imaginative vocal tone with exaggerated musical inflection to convey clear, emotion-driven signals. Here I believe we have the essence of tonal-musical-emotional expression.

Teachers and singers talk of *colouring* the voice, but this is not something we *do* to it. The singer's *aural* palette must become as sophisticated as an artist's visual palette. Stimulated and fed by text, language, musical nuance and sentiments in accordance with his understanding, a great range of colours and shades is called forth from the singer's throat without forethought. In an imaginative singer with a free voice, 'colours' mix as if by magic, since *what* produces them is the very stuff of the musical-emotional tone-stream.

As our singing voice becomes freer and we regain our aural familiarity with it, we discover that tonal and musical imagination combine naturally to serve the verbal imagery hatched by words. This is the same kind of merging as when words combine with emotions to swing the speaking voice out to and beyond the limits of its normal range. These 'alchemical' mergings are not contrived; they are indications of freedom in an imaginative and intelligent singer.

Aural imagination plays a crucial role in 'leading us back' to our singing voice. All that chortling and bird-like experimentation we made 'discovering our voices' as tiny children was spontaneous, without instruction, and made with increasing delight and variation as it became more conscious. All too quickly, however, what produced these sounds was commandeered for speech. Through months and years of determined intellectual service, the natural lyricism of the voice 'gives up' for want of stimulation and practice, leaving us with sounds which are no longer vocal in the *singing* sense.

To better understand the nature of aural-vocal connection, we could imagine our singing voice as a picture of flowing shape and form, composed of tone colours instead of pigments. After years of neglect, and the assiduous 'imprinting' of speech patterns upon it, the true shape, form and colours of the original can no longer be made out. It's in this sense that we've gone deaf, or 'aurally blind'.

While our singing voice as a whole cannot be *seen* to be working, nor its complexity physically felt, and although the true picture of our voice may be obscured or half-forgotten, our ear seems to retain a *memory*, a blueprint or complete picture which, with due care and thoroughness, can be revealed. Not only have our ears the capacity to hear our singing voice in its entirety, but (with the help of our imagination) they can play a crucial role in their own reawakening process. Our ears are an indispensable 'observer' and 'recorder' as we re-explore our unique vocal world.

All sounds we make with our voice reflect what it is that produces them. This is of course physical, but, because it is at times an intricate and invisible matter, the teacher must learn to recognise and assess the activity *aurally* rather than visually. Similarly, as the physical work progresses under the teacher's guidance, the ear of the singer, in *recognising* the original form of his voice, resumes its role of vocal monitor.

As the voice regains its natural balance and strength, not only do we feel less inclined to control it physically, but the desire to 'listen' to it in our normal conscious thinking mode recedes. A paradox? Not at all. We never bothered about *how* to express ourselves vocally before we learned how to do so verbally, and were pretty good at getting our 'message' across without explanation. In the process of recognising and restoring the physical voice, we reverse the aural decline by re-establishing the aural-vocal lines of communication, thereby obviating the perceived need for a more tangible way of judging and controlling our voice.

Vibrations elicited by vocalisation can be felt in various localities around the larynx, which are commonly referred to as resonators or resonating cavities. While there can be little doubt that these vibrations affect our vocal hearing, in searching for such vibrations we are just as likely to unbalance the voice as when we look for other types of physical sensation elsewhere in the body. In a balanced voice, physical sensations are spread throughout the whole vocal complex and are felt only when exaggerated, as in various training processes.

In the normal voice, which cannot sing, there is no neurological or physical impediment as such. The reawakening of our singing voice is a neuro-physical realignment, brought about by appropriate physical readjustment – a change of vocal-aural habit. Once this is achieved, we cease the endless analysing to which singers in general have become so prone. When young, we were never aware in the intellectual sense of the particulars of our voice. Our aural-vocal 'knowledge' was something with which we were born (pre-reason) and so tied up with our being that we enjoyed it and played with it, only knowing that it existed. Its existence was our existence, and we 'knew', without consciously knowing, that we were alive.

In the following three chapters I shall explore three aspects of human development that have contributed in significant ways to the diminishing of our ability to sing our song:

1. Adverse effects of learning to speak;
2. The repression of emotion;
3. A sedentary life-style.

These are consequences of civilisation. To civilise means 'to bring out of barbarism, enlighten, refine and educate'.[1.4] Laudable aims, but we might usefully ask to what extent we have actually achieved any of them, and (whether we long to sing or not) at what price.

PART I

CHAPTER 2

Sounds Intelligible

While humans are predisposed to socialise, we are not genetically programmed to speak any particular language, not even our mother tongue. The human need to communicate is nevertheless deep-rooted, and it seems likely that a group of infants unexposed to any existing language would eventually develop a new one out of their baby babble and vocal-physical gesturing.

Intelligent strangers

If our hypothetical group of babies were split into two, it is reasonable to suppose that each group would develop a *different* language, and the groups would thereby become strangers. In spite of the language of each group being incomprehensible to the other, both would have been inspired by the self-same human motive and a vocal instrument of exactly the same construction. The vocal *communicating possibilities,* therefore, are shared at the outset.

The voice with which we were all born, which has the *capacity to unite us* through the expression of universally shared emotions as well as the shaping and sounding of words (which are in turn responsible for the flowering of a *shared* intelligence) in a sense separates whole groups of our human race and even individuals within the same group from each other. Furthermore, for all its importance, the acquisition of language speaking will be seen to divide us from ourselves.

In some quarters, sign language is held to be as expressive as voiced language. The advantages of voiced language, however, would appear to outweigh those of signed language in several significant respects. The most obvious is when the hands need to be otherwise employed while speaking. Two other characteristics, exclusive to voiced language, command our attention: natural carrying power and its roots in emotional utterance. Both these characteristics are amply demonstrated in young children, but seem to become increasingly redundant as we grow up.

The voice that carries without effort attracts. We were all born with this voice, and can therefore assume that it was intended by nature. Indeed, we might reasonably deduce that this easily-delivered, attracting voice was part of an efficient communications plan. Nowadays, communication over any distance usually requires technical assistance if it is not to sound harsh or strained. This further distances the speaker from the listener. Someone who speaks with a clearly projected voice, and is thereby always audible is often considered impolite or embarrassing. For want of practice or encouragement our *original* voice loses its ease and power of projection, setting up a potential barrier to communication.

Travelling light

When we cannot for some reason express our feelings we feel blocked, heavy or impotent. Our feelings being part of who we are, we suffer if we cannot voice them. *Expressing* what moves or hurts us is healthy because, rather than denying the truth, we are accepting it and 'getting it off our chest'.

Poets and authors verbalise emotions to give us a way of 'beholding the truth' even if we cannot otherwise look it in the eye. As a singer you'll not be unaware that the majority of songs are at least tinged with sadness: parting, unrequited love, nostalgic longing and so on. Song recitals could be depressing affairs but instead are a kind of *celebration.* The depths of our heart's sorrow as well as the heights of its joy are lauded in poetry and the music it inspires. Instead of avoiding these sentiments we jump at the opportunity to express them in song. This is no self-indulgence; as well as a celebration, singing is a *cure,* a lightening of the spirit and as necessary for its health as food is for the body.

In speaking, what is heard is a relatively feeble or indistinct version of what is felt. Even with our best efforts, our tone of voice tends to be obscure or our emotional message garbled. It's easy to be saying one thing while feeling another, and our emotions don't easily give us away, any more than we give them away. With singing, on the other hand, we are given *carte blanche* to air feelings ('borrowed' as they may often be) which otherwise we're inclined to hide. We may express the otherwise-for-us inexpressible and share our uncensored humanity with and on behalf of strangers. We should not be surprised to discover then that, unlike the speaking voice, the singing voice is equipped to express clearly and directly in sound the whole gamut of human emotions.

Hijacked voice

When our intelligence began to take root it found a ready partner in the versatile vocal equipment with which we were already endowed. Looking at the word 'language', we see that it shares the same root as 'tongue'. [2.1] Looking at the voice more closely, we realise that the emission of vocal sound as such does not require a tongue. Apart from chewing, sucking and swallowing, the tongue is needed only to give our sound intelligibility – intellectual meaning.

The tongue satisfies both our stomachs and our intellect. Designed with vegetation in mind, the tongue initially aided chewing and swallowing manoeuvres, and lolled about like that of any quadruped. When we eventually stood up to get a better view of the world, instead of making adjustments like other parts of the oral cavity, it settled back in our mouth, taking up a disproportionate amount of space. Singers would have no problem if all had continued to be well when with the tongue's aid we began to invent language. Alas! All did not continue to be well: the beleaguered tongue finds itself in the dual role of hero and unwitting villain. It is a hero because it enables us to speak, a villain because in singing we often get so 'tongue-tied'.

With a few simple experiments we can experience where the voice-language conflict is likely to have begun.

Experiment 2.1

On a comfortable note sing the following two-bar phrase:

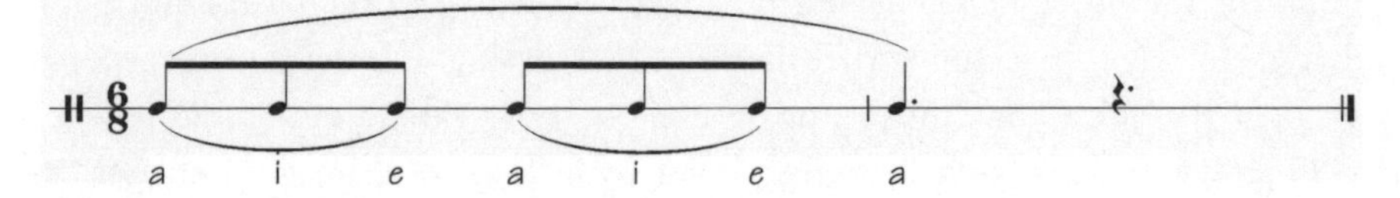

The sung tone, which in singing should be confident and continuous from start to finish, is 'changed in shape' thanks to the tongue, and only the tongue. The jaw and lips should remain still and relaxed throughout. Try this at different speeds.

The success of this experiment depends on two basic but separate ingredients: the freely given but stable stream of tone, and the free but concise movement of the tongue. Neither need interfere with the other.

Experiment 2.2

Firmly closing (but not 'clamping') your teeth together, try the two-bar phrase again. Remember the unbroken stream of tone, one impulse only to see you through.

Experiment 2.3

Hardly opening your mouth, sing babibe/babibe/ba or dadide/dadide/da or use 'm' or 'p' or 't'. I suggest you do this quickly rather than slowly. Neither the tip of your tongue nor your lips should be lazy!

If you were unable to maintain the constant tone in these 'experiments', it meant that the way you articulate words adversely influences the way you produce your voice. Likewise, if the tone was constant, but clear verbal articulation was difficult to achieve, it meant that what should have been free in the enunciation department was to some extent occupied in 'assisting' the emission of the sound – the tone was *not* freely emitted after all.

Unless we acknowledge that the singing sound came into being before the invention of language, we are in danger of confusing the issues of voice emission and the articulation of text, instead of benefiting from their partnership. Unless the singing voice is emitted according to its original design, and unless text is articulated in a free and efficient fashion, text will always threaten the tone stream's integrity.

Voicing text

The only way that voice and sung text can truly complement one another is if both are freely achieved. Interestingly, good verbal enunciation in speaking doesn't necessarily make for good vocal emission, whereas a freely emitted voice does pave the way for perfect enunciation. The natural *singing sound* is primary; it does not need the help of the tongue or jaw or lips, which are free to go their verbal way. Speech, on the other hand, having deranged the throat's original natural vocal activity, is in no position to instruct the voice how to sing. On the contrary, speech has taught the voice how not to sing. When we try to extend our speaking voice into singing, a vicious circle is created, whereby lack of vocal freedom obliges the tongue or other swallowing or jaw-operating muscles to 'assist' the voice. So long as the tongue and jaw are even partially employed in this manner they cannot do their job without undue effort. Attempts to simply improve poor enunciation usually lead to misplaced effort, and time is wasted. Furthermore, the vocal problems which lie behind this poor enunciation are often exacerbated.

Any singer with an adequately free voice knows that in a sequence of sung vowels (such as a-e-i-e-a), the movements required of the tongue are small, as its sole job is to slightly alter the shape of the space in the mouth. With 'a-o-u-o-a', lip movements can be similarly economical. The stream of sung tone is a constant at its source (the larynx), and then is 'shaped' as it passes through the mouth. We have called these shapes a, e, i, o and u. The unimpeded tongue and lips can make these and hundreds of other shapes precisely and fluently. With this facility, subtle changes of vowel colour, necessary for a wide range of sung languages, can be made without undue effort or tension. The same can be said for the lips, teeth and tongue tip when it comes to articulating consonants.

Babies, for whom words don't yet exist, vocalise in an increasingly conscious way, using their voices in a 'singing' manner, smoothly making random shapes of sound, which, melodic and vowel-like, are neither music nor language. While their voices are unimpeded, their tongues taste, suck and swallow, and their lips keep busy – free and competent activity, just as it was designed to be. Here we have both the beginnings of sung tone and the ingredients for articulate speech.

Unlike walking, speech needs to be taught. Normally a baby imitates her mother, beginning with something simple like 'mama'. She has no difficulty combining voice and lips; on the contrary, she gets the sound quicker than its meaning, and 'mama' develops delightedly into mamamamamamamama….*ad infinitum*. 'Baba' and 'Papa' are just as easy, while in French 'Maman' and 'bébé' are hardly more difficult. (Notice in passing that when reverting to so-called 'baby talk' the adult tongue is happy to rest!) In English in particular we have the potential problem of diphthongs as in 'baby' (bei-bi), but a baby's free explorations already inadvertently include various elisions – further evidence that the voice we're born with lends itself readily to the development of speech.

Faulty transmission

There are clear risks that go with the process of learning to speak. If a baby's teacher has a weak, breathy, harsh, nasal, or otherwise blocked vocal emission, manifesting vocal defects or inefficiency, she'll pick these up to some degree in her speech learning, effectively becoming vocally defective or inefficient herself. Remember, speech is facilitated by two distinctly separate systems, which appear to be able to complement one another: that which produces vocal sound (sheer voice – larynx), and that which shapes it into recognisable language (articulatory mechanism – mouth). With bad voicing on the one hand or poorly shaped language on the other, complement turns into conflict.

Voiced labour

Learning to speak, through no fault of its own, has become a laborious, ungainly process, with unfortunate consequences for human communication. At what point in our evolution the balance between emotional and intellectual expression tilted in favour of the intellect we can only speculate, but from the vocal standpoint it's clear that it did so. This process continues with the development of each child who reins in his feelings as he becomes more rational and verbally articulate. Poorly voiced and poorly articulated speech has become the norm. When speaking becomes a serious business, as, for example, in acting or teaching, weaknesses and inefficiencies begin to tell. The spawning of electronic communicating equipment has added laziness to the problem. Considering how incompetent our voices have generally become, it's a wonder that the human species has not altogether lost its capacity to sing. For all its complexity, normal speaking employs very little of the available vocal apparatus (sufficient to form bare words, often rather monotonously). The rest goes to sleep.

Split personality

The mind-body split created in much speech is nowhere more evident than in our efforts to sing. Perhaps this is a vital clue to the general desire to do it: do we miss the true expression of our emotional life?

Two things at least seem clear:

1. We have something more than words to voice, and more than our intellect to satisfy.
2. The formation of words represents the least of our vocal capacities.

I have come across a number of performers in other fields (dancers, instrumentalists and speakers among them) who, although already expressing themselves very well, would really like to be able to sing. I've met solicitors and school teachers, plasterers and plumbers, farm and factory workers, who say they'd love to sing. Work songs, folk songs, and religious songs all bear witness to the human need to voice and honour the human spirit and to celebrate human life. The singing voice enables us to go beyond words and to reach deeper into the human condition.

We are all born with the *capacity* to sing. When we learn to speak, and when we refuse our feelings their rightful voice, we unwittingly reduce this capacity – a kind of self-throttling process – until we have the urge to find our voice again, or it attempts to surface. For some their singing voice seems irretrievable, for others almost within reach. A lucky few appear to have withstood the negative effects of civilisation that I'm describing here. These are usually referred to as 'natural singers'.

Confusion

Although we can only speculate as to when human beings became talkers rather than singers it's relatively easy to see *what* has happened, *how* and even to some extent *why*. The voice and 'word shapers' started out as parallel and complementary paths, ready and willing to serve a mind eager to develop itself. Gradually they converged, becoming less distinct and more confused, as many subtle linguistic devices took over the vocal system. Vocal 'distortions' crept in – aspirates, guttural and nasal sounds, diphthongs and multiple consonants – anything of which the vocal tract was capable. Some parts of the vocal mechanism were favoured over others, leading to vocal one-sidedness, incapacity to project the voice, and, even more seriously, emotional dislocation. Although the underlying component parts (voice and word-shapers) are still there, they are now generally no longer true either to themselves or to each other. In dominating the situation, word-shapers become voice *re*-shapers. In other words, the pure voice (the singing voice) instead of complementing language and lending it mechanical ease, emotional depth and character has become subservient to it, even to the extent of giving up. So long as we use muscle systems (as long as we *have* a use for them) they have a good chance of remaining in working order. Neglected, they deteriorate rapidly.

We human beings are talkative, often to such an extent that it seems we need the sound of our voice to reassure us that we exist! If we so often talk and so rarely sing, and there is such a comprehensive difference in muscle use between these activities, it is hardly surprising that singing seems a rare occupation. The impression that the *ability* to sing is a gift granted to a few, while the rest of humanity is in this respect voiceless (and by extension has no music in them) is as misleading as it is unfortunate. In relegating our voice we lose authority and authenticity; we lose psychological power to assert ourselves, physical power to call or summon from a great distance, emotional power to bridge the voids between hearts. While it's unfair to suggest that learning to speak alone is responsible for separating us from our true voice, it is easy to point to evidence that it has played, and continues to play, a major role in this self-dividing process.

Laryngeal collapse

Experiment 2.4

Take a story, and read it out loud, as though to somebody, but read on and on without taking a breath!

You may be surprised how far into the story you can get on one normal breath. However, this experiment really begins at the point where you feel you have run out of breath. At *this* point (ignoring the strong desire to inhale) encourage your larynx to go on at least for a few more seconds. (***Caution:*** If you are over-enthusiastic or too insistent you might panic through lack of oxygen or hurt your throat; this is not an exercise!)

What happened to your larynx at the end of this bizarre experiment? Probably it felt tight, no doubt it sounded it. As you read, and expended your breath beyond the point at which you would normally replenish it, your larynx experienced greater and greater isolation; in other words it lost any remnants of 'support' it might have had in its familiar speaking mode. In spite of our diaphragm's increasing urge to inhale, the muscles designed to support the larynx collapsed. It was this that created the opposite sensation to that of an 'open throat'.

Laryngeal support

As we learn to talk and emotional expression is discouraged, a gradual reduction of natural laryngeal support takes place unnoticed in almost all normal throats. Normal speaking does not require great strength in the extrinsic muscles of the larynx. On the contrary we can and do get by with just sufficient to prevent that 'scrunched up' feeling experienced at the end of Experiment 2.4.

Conversely, the more these supporting muscles are re-invigorated and strengthened, the greater operational freedom the larynx gains, until it is restored to its open-throated singing condition. We need to know, in teaching and singing, as well as in unusually demanding speaking, that the larynx requires this stronger, more complete support, and we will see later why efforts to 'support the voice' might better begin in the throat than the thorax.

Dynamic tension

The sounds that we most often describe as vocal are those most easily discernible from the larynx itself – the sharper, more clearly defined variety. These result in the main from the *intrinsic* laryngeal musculature which is responsible for creating tension in the vocal folds and bringing them together. Through their combined action, the amount of air passing through the glottis is reduced, thereby concentrating the sound. Simultaneously, as the strength of this system increases (and every pair of vocal folds has its natural, individual tensing capacity), greater support is demanded of the *extrinsic* muscles, which, so to speak, 'take the strain'. Tense and close the folds without adequate support from these muscles and the voice gets tight. Weaken the vocal fold tensing or glottal closure and the sound becomes weak or breathy. Tightness, breathiness and lack of tonal clarity are all typical of normal speech. Well suspended, the larynx is incapable of making harsh or strained sounds but acquires warmth and sweetness without losing incisiveness.

Compensation

In order to forestall the discomfort that can easily be experienced by the 'normal' throat in attempting to sing, various other non-singing muscles (such as the tongue and other swallowing muscles) volunteer their services. These muscles are compensating for those which are too weak to do their natural larynx-supporting work. In voice training, it must be remembered that these compensatory tensions are there for a reason, and will only be satisfactorily relieved when they have been rendered unnecessary. It is only when the natural suspensory system is fully operational that the throat is free of unhealthy tension, and can intone and 'speak' its message simultaneously.

The emotional connection

We are now poised to make a significant connection between our voice (insofar as that means the larynx) and our emotional expression (in so far as this is a physical matter). It is important to get this right, because it has as great a bearing on vocal balance and health as anything. It seems odd that anyone dealing with the human singing voice can think of emotional expression as optional. It is necessary, not simply because an emotionless voice is a dead one, but because we cannot still the diaphragm or block emotion without severely disabling the whole vocal system.

The diaphragm is a large muscle, being the main muscle of inhalation. For the air to be drawn into our lungs with ease, the throat – more specifically the glottis – must open. There is, therefore, a direct connection between the opening of the glottis and the diaphragm, and, crucially, it is the flexible support system of the larynx that links the two.

Breaking the bounds of the mind

The collapse of the suspensory mechanism, which has effectively 'disfavoured' the emotional content so natural to the human voice, has confined the inflection and modulation of the speaking voice to a relatively narrow compass. If we venture beyond this normal speaking compass we involuntarily make a kind of 'vocal gear change'. Venturing higher in pitch, we slip into a sound commonly referred to as 'head voice', 'head register' (reflecting the vibration of lengthened vocal folds) or *falsetto* (the difference will be explained later). At the other extreme we change into 'chest voice' or 'chest register' (reflecting the vibration of tensing vocal folds). These modes are often cultivated for effect: the one can sound 'feminine', the other 'manly'. It's only with abnormal emotional excitement – in flights of hilarity or explosions of indignation, for example – that our voices reach out beyond their relatively narrow compass, thus breaking the confines of speech.

Passaggi

The vocal zone in between these extremes, the so-called 'middle voice' or 'middle register' (the range in which we do most of our speaking), which in normal circumstances is quite comfortable, is often thought to be the one most disposed towards singing. Believing this lends credence to the idea that singing is an 'extension' of speaking, and that to increase our singing range we should set about building bridges or 'links' to the outer tonal registers.

The neither-here-nor-there zones between registers are historically and popularly described as '*passaggi*'. 'Registers' and the weak areas between them are generally viewed as natural, thus lending weight to the argument that voices cannot be even in tone from top to bottom. *Passaggi* are the justification for entire techniques devoted to bridging processes, in which the general idea seems to be to concentrate on the middle voice and 'work outwards'. This appears to be logical until we realise that not only are the outer registers weak through lack of use, but that our middle one is normally in a state of collapse. This zone will turn out to be the one that most commands our attention, because what in *singing* terms characterises it as a 'legitimate vocal register' has, in most voices, ceased to function. Generally, while the head register misses the chest register and vice versa, and both miss the middle, the middle register misses what it needs from the outer ones, and lacks its own specific vocal identity. This state of affairs is instructive for voice training since it indicates working from the outer zones if we want them to contribute in any significant way to each other's welfare, and at the same time re-awakening the middle zone to discover its crucial 'mixing' role for the whole voice.

'Register break' is perhaps a more appropriate term than *passaggio*, because registers are the manifestation of operational breakdown – one muscular configuration in the larynx giving way to another. This can be clearly simulated by yodelling (deliberately breaking across the zones) or 'breaking' from the head voice (or *falsetto*) to the chest register or vice versa, stretching to tensing and back again. (Try yodelling for yourself. Unforced, the larynx likes to move in this way!)

While the middle register applies to the activity of certain muscles without which no true singing or 'mixed' sound (*messa di voce)* can be achieved, its somewhat refined activity is nowhere to be heard in normal speaking. It is *these* muscles that turn out to be the most difficult to arouse from sleep!

Only for singing

Perhaps the most striking aspect of the human voice is that the source of the singing sound is so complex. As part of their structure, the vocal folds have a system of finely calibrated muscle-bundles and muscle-fibres, which radiate to their outermost margins. Furthermore, as Robert Sataloff writes, 'The vibratory margin of the vocal folds is much more complicated than simply mucosa overlying muscle. It consists of five layers…'.[2.2] At the very least this obliges us to view the human voice as a *precision instrument.* It is the muscular structure underlying the mucosa that is responsible for coordinating the most intricate lightening-speed adjustments between the stretching and tensing processes of the vocal folds, and facilitates and controls the qualities which most characterise great singing, among them infinite variations in tonal colour and substance and every degree of musical and emotional nuance.

Even though in the majority of us this 'edge mechanism'[2.3] has gone to sleep or has atrophied through neglect, and is therefore apparently redundant, it constitutes the heart of a highly original instrument. Rooted in the base of the folds and reaching with increasing complexity to the very edge of human utterance, this sophisticated human attribute is used for nothing but singing. Should we ignore such a clue to what singing may be about?

Divine design

Not least of the far-reaching implications of this intriguing fact is that we human beings were *designed to sing.* To quote Husler and Rodd-Marling, 'the edges of the vocal folds divide into harmonically ordered sections which can have no purpose other than to produce "useless" aesthetic sensations.'[2.4] We are first and foremost, in the deepest and most vital sense, *singing beings*, and only secondarily speaking beings. In other words, it is the singing voice which connects most intimately with and expresses most truly the essence of our humanity, our 'soul'. It is the singing voice – whether we sing or not – which most fully and clearly defines us in sound. We have a phenomenal instrument of great power, beauty and refinement – for what? Did the Divine Creator anticipate by millions of years that one day some lesser gods (Monteverdi, Mozart, Verdi, for example) might need such an instrument, building in this refinement with the possibility that eventually people might want to invent something called *bel canto*? At what point in our evolution, prior to the invention of music as we know it, did we gain (or indeed begin to lose or reject) this astonishing human attribute?

There seems to be a paradox here, which becomes more striking as we consider what these tiny muscles signify in singing terms. The term *messa di voce* ('mixed voice', not to be confused with *mezza voce,* literally 'half voice') implies that the voice can also be *not* mixed. As we have seen and can clearly hear, the normal state of vocal affairs is one of dis-unification, head voice seemingly somewhere up in the head, chest voice down in the chest, and in the middle, seemingly much nearer to the throat, a somewhat amorphous sound which we call the speaking voice. Because of not keeping up our singing beyond a few months of life, parts of the instrument suffer neglect, going to sleep, splitting off from the whole or barely surviving at the extremities of our vocal range. This is a vicious circle: we don't sing, the edge mechanism languishes, we lose our vocal *integrity* (our singing voice) and thus no longer sing. Is it only the ability to sing that we lose?

Messa di voce

The weakness of the muscles (or their lack of response) goes a very long way to explaining why voices are generally uneven, unstable or divided into unrelated parts. In a healthy singing voice, the marginal muscle-fibres interlink two basic antagonistic vocal activities – the tensing and the stretching of the vocal folds. In understanding this we see that their activity is not a question of pitch; they play an indispensable part in maintaining the integrity of the vocal structure at *any pitch*. Their effect is to ensure tonal continuity in a way that eludes the most ingenious *passaggi*-bridging strategies. It might be more accurate, as well as more useful, to see these muscles as a kind of physical-tonal catalyst, since they facilitate tonal and dynamic variation without it disturbing overall structure. Another useful analogy for the edge mechanism would be that it provides a kind of physiological 'linchpin', ensuring the finest anatomical organisation throughout a singer's natural vocal range. In any case, the sound world that a truly mixed voice brings about far surpasses the sum of its various anatomical or tonal parts, call them registers or what you will.

The integrating factor is of vital importance. For a start, this specialised system brings about the instrumental, tonal and expressive refinements coveted and prized by all sensitive singers. Then, in bringing about optimum vocal balance and intensity, it is responsible for the *wholeness* of our human sound. By this, I don't mean merely abstract musical or tonal quality. In every sense at the centre of vocal events, this integrating-refining system is able to reflect the very core of our being. It's instructive that this 'heart' of our voice is what we must work most assiduously to regain and nurture if we wish truly to sing our song.

PART I

CHAPTER 3

The Stifled Cry

Our many emotions – such as tenderness, awe, agitation, sadness, fear – can be quite definite, or indistinct. We may react to emotions, act upon them or try to put them into words, but they tend to come and go as they please. Although it may be obvious that various physical and chemical changes occur as one feeling replaces or mingles with another, it is not easy to describe in physical terms what exactly emotions are. What is irrefutable is that we all have them, constantly and in abundance, and scientists now accept that such elusive 'movers and shakers' are worth exploring.

We have various means of expressing our emotional life, some active (such as caressing a loved one, or banging a fist on the table) and others passive (facial expressions and other involuntary tensions). When it comes to the voice, we might sigh or shout, or else put our feelings into words, a 'mental framework'. However, it seems to be a mark of our modern civilisation not to express how we feel (or even what we think) in a clear, direct manner. Instead, we must be cautious, polite, and politically correct. What we say in fact often belies what we're *feeling*. We prefer to say what we *think*, or what we want others to hear us say rather than expressing our true, raw feelings, or else we 'explain them away'. This is exacerbated by an inability to express our emotions vocally. For various reasons some of the physical mechanisms involved have become inhibited, or dysfunctional. Tone of voice is usually confined to the vocal limitations of speech, lacking focus, range and colour. These limits are only overcome when emotion is allowed to well up, or is shocked into coming from a deeper or higher place than usual. Then, the connections between the source of feeling and the source of sound make a more complete, effective partnership, for example in the expression of extreme despair or of righteous indignation, as well as the sing-song conversation we employ in relating to infants.

Our tone of spoken voice rarely reflects even our most heartfelt feelings with clarity. Emphasis or affirmation may come out as merely loud. Anger and pain get stuck in our throat, so that their expression is blocked. Having learned not to be vocally expressive, timid attempts at being so may sound mumbled or feeble. If we are emotionally vocal at all, we betray a *struggle* to express ourselves. Even when we find our voice, metaphorically or literally, social norms and the habits of a lifetime can persist in modifying or mollifying our grievances and even subduing our joys.

Being able to sing doesn't mean that we'll always express our feelings with spontaneity, forthrightness and clarity, but it does open up the possibility of doing so. Furthermore, in opening up the vocal-emotional channel we connect to ourselves in specific physical-emotional ways. This not only increases our ability to vocalise our feelings but connects us to emotions with which hitherto we were out of touch. The voice can act as a *self-revealing* system. This is enormously significant for singing, not only in the sense of enabling a singer to be faithful to prescribed sentiments, but in the extent to which his voice is emotionally sensitive and spontaneously creative, making the crucial difference in performance between something soundly reproduced (judged to be 'good' for various specific reasons), and something which places vocal communication on an altogether deeper level.

If we think that expressing feelings in singing means losing control and being overwhelmed by emotions, emotive singing will seem distasteful and unmanageable. Feelings in singing cannot be faked, or simply 'added on'. Unless fully integrated physically, they sound unconvincing, embarrassing or ridiculous. On the other hand, attempts to be expressive without emotion usually sound inappropriately cool, 'contrived', impersonal, or lacking personality. True emotion, for which our singing voice is a natural outlet, does nothing but enhance singing, making it more believable. Fully integrated, emotion does not threaten the voice's stability but enlivens it, giving it genuine humanity. True singing is grounded, and avoids hysteria by manifesting genuine strength. The humanity of a singing voice expressing genuine emotion explains why certain performances can be so moving and compelling, much more than an intellectual or aesthetic experience.

Transformation

The power of singing lies largely in its transforming qualities: the ability to move *the other*, the fellow human, into being, into feelings that otherwise might remain hidden. Singing can help others (especially those who cannot sing) to experience their song through a kind of empathy. Singing can help people to break out of their emotional constraints and feel truly themselves. Those who are fortunate enough to find their voices therefore have a responsibility to those who are not.

Physiology

In comparing the physiology of the singing voice with that of voiced emotional expression, we discover that they are, to all intents and purposes, the same. In fact the expression of emotion and giving voice both arise from the brain's endorphin system.[3.1] So, in denying or suppressing emotions in singing, we are in danger of inhibiting or restricting the voice itself. Conversely, by encouraging the vocalisation of emotion we can stimulate a voice into being true to its expressive nature. Singers often find that their voice and their emotions are in conflict, so, before we look in more detail at how our singing voice and our emotional physiology work as one, it may be helpful to examine what commonly happens to separate our feelings from our voice – to render emotions voice-less.

Demons and Diamonds

Expressing emotions – giving voice to the excitement of our feelings – is not necessarily a question of will or propriety. As often as they creep up on us without our bidding, they seem to be remote or locked away. Carl Jung, whose investigations into the human subconscious were profoundly revealing, urged us to explore, come to terms with and even embrace what he called our 'shadow'.[3.2] By this, he meant the parts of our personality and human nature which we have suppressed, denied or hidden away in reaction to trauma, physical or psychological abuse or as a result of years of cultural conditioning. As young children, we often suffer injustice or humiliation, or are made to feel impotent. Because these experiences insult and wound our spirits, we devise (subconsciously) ways of avoiding the re-experiencing or re-opening of these wounds. If, however, a wound hasn't properly healed, it can affect us in surprising ways. The physician Sir William Osler said 'The hurt that does not find its expression through tears may cause other organs to weep'.[3.3]

Painful tricks

The rejection or denial of hurt, as necessary as it might feel at the time of abuse, can play tricks on us, or wreak havoc with our behaviour. For example, we might get angry with someone because their behaviour *reminds* us (subconsciously) of something which, perhaps at an age when we could not or dared not express it, caused us such suffering that we consigned it to oblivion. To defuse this anger, or to even find a positive outlet for it, we must learn something about its origin, so that it's no longer potentially destructive. Likewise, if our needs, particularly the emotional ones, were ignored or barely met in our past, we may suffer feelings of emptiness or isolation, or of being of no value. It's not difficult to see how such feelings might affect our performance on and off stage.

To be fully in touch with ourselves, fully authentic, we must pluck up courage and rediscover the hurt parts of our being and take care of them. We may need help with this, but so long as we shirk this 'trial', we continue to be defensive, against others and against ourselves. For singers this can be particularly significant, because of the physical manifestations of emotional defence which can make it difficult either to give or to be received when we sing.

Pain past and present

To fully realise the implications of this self-blocking, we must understand that, although its cause may have long passed, original pain or insults have not gone away. Physically as well as psychologically we hold pain at bay, keeping ourselves from falling apart – or so it seems. Unfortunately, although the original experience is a subconscious memory, emotions connected to it have a nasty way of resurfacing in different guises. Ongoing suffering, even if no longer consciously felt, can undermine our performance as human beings generally. So long as singers remain emotionally 'protected' their physical communication in sound remains to some degree forced, untrue or apologetic.

This muting of the audible voice of the body and of its emotional life diminishes the ability to convey in sound the deeper reaches of our being in anything but relatively lifeless words. Even with our best intentions, our emotions often have a struggle to be heard, let alone understood. If your capacity for vocally expressing anger, for example, seems to have been stifled, your voice might sound deceptively soft-grained, lacking bite or focus. Although we cannot assume that all singing of this description masks an angry person, neither should we be seduced into believing that the gentle wisps of vapour issuing from a harmless seeming volcano signal that all is well beneath the surface.

Anger, like fire, can be turned into a galvanising agent or creative force. If you are no longer its subject, you can return to it *knowingly*, and find yourself with a power that you can turn to good use. In singing, the 'angry' part of the voice is what lends it its inherent strength, its authority, its dramatic potency. This demon, among others, when handled with knowledge and due care, can be positive, a diamond after all!

Diamonds

Perhaps we should not be surprised to discover that the shadow also conceals buried treasure. Less obvious than thwarted anger but just as potent is negated happiness. Take a happy or exuberant child, and constantly command her to keep quiet or still whenever she spontaneously shouts or dances for joy. The message she gets is clear: 'It's wrong to be overtly happy'. Worse still, the subtext seems to say 'It's wrong to *have* such feelings'. The natural sparkle of bright light is dimmed, perhaps altogether extinguished, and replaced by a pall of sadness and resentment. The child's joyous response to life is superseded by the questioning of feelings per se.

If locking away our feelings was a rational process, it might take only a key of reasoning to reverse it. But a young child does not think 'now I'm happy, now I'm sad'. She does not reason, 'I am not being allowed to be angry or ecstatic'. Her feelings and their spontaneous expression are perfectly natural responses to life. It is the *body* that laughs and dances in delight or cries in anguish; it is in this physical expression that the spirit is satisfied. An insulted body restrains its own life; and the spirit, refused its natural means of expression, languishes.

Guilt and shame

The fury and frustration, anguish or loneliness which make us feel inadequate, marginalised, defiant or desirous to please can be exacerbated by self-blame or feelings of guilt or shame. However, the clinical psychologist Dr. David Smail once wrote of those drawn to or referred to psychotherapy, 'They are less people with whom anything is wrong than people who have suffered wrong'.[3.4] Once we are adult, we must take responsibility for how we are. The starting point of becoming what we were *intended* to be or to become (as distinct from what we have been conditioned to become) is the *acceptance* of ourselves as we are, warts and all. In accepting ourselves we can admit that we need help. We are *designed* to laugh and cry; to transmit our emotions physically, to convey how we feel *vocally.* Discouragement, censoring and prohibition of emotions debilitate the expressive reflexes and physical mechanisms with which we were born. This causes chronic retentive or protective muscular tension.

There can be few people who do not have demons of one kind or another lurking in the shadows, or who do not have diamonds which only need the light of day to sparkle and show their worth. The process of learning to sing has a way of uncovering or releasing the parts of our personality which are dormant or afraid. This can be an exciting 'growing up' process; it can also be painful. Time and effort is sometimes wasted in the studio because there seem to be so many doors, or too many locks. As teachers our job is *not* that of a psychotherapist. We must keep an eye or ear open for signs of psychological distress which may indicate some other form of treatment.

Our bodies and emotions are in league with one another, each in its way reflecting the condition of the other. As well as taming the demons and discovering how they can serve you, singing should be a search for those diamonds, those life-affirming aspects of our personality which as children we carried determinedly but lightly into this world. Successful therapy or healing is bound to involve the whole person, body, mind and spirit, which is why the holistic process we call learning to sing can be so liberating.

PART I

CHAPTER 4

Bowing to Life

We have already begun to examine certain areas of the human structural design. All bodies operate in the same fashion and to some degree everyone can cultivate the same physical skills, such as dancing or typing. All human beings are born with the same vocal structure which has the potential to sing as well as to speak.

Our skeleton, comprised of rigid pieces connected by ligaments, resembles the armature of a clay sculpture, except that the human armature is articulated. While our bones' structure determines our overall form, we would be mistaken if we thought that our skeleton took care of itself or that *it* kept *us* in shape. Our bones are not rigidly glued together any more than they are precariously balanced one on top of another; we are articulate in every joint, so that we can move about with ease. But why don't we fall over while continually moving around in ways that seem to defy gravity? In all our movements there's an engagement of opposite forces, and a reflexive adjustment of weight, so that not only can we move easily back and forth by minute degrees but exactly and instantly gauge the extent and strength of our movements, depending on what we want our body to do. Add to this the balancing sensors housed in our ears which register the position of our head in relation to the earth, and our neuro-muscular (proprioceptive) system in general, and we appear to have all the equipment we need to prevent us from losing our balance. However, the head is heavy and perches on a tapering spine, and gravity is a force to be reckoned with. Is it really by means of muscles that we hold our heads high?

In talking about posture we're not referring only to standing or sitting, but to the constant relative re-positioning of the various mobile parts of the body in *any* activity. We are concerned with ergonomics, with physical efficiency, which in modern times is often threatened by non-conducive circumstances.

In everyday living we seldom need to consider which muscles are doing what, or how they are managing to do anything. While we're not puppets, our muscles do seem to work by a 'pulling of strings'. We can explain this with neurophysiology and biochemistry, but to what extent can we assume that everything muscular is duly taken care of? How much *control* do we have over how or how well we move?

Self determination

We normally attribute our individual physical peculiarities, at least the observable ones, to our genetic makeup. Whether we are of stocky or slender build, for example, is pre-determined in the mix of genes inherited from our parents. There are other factors in life which, regardless of our genetic make-up can influence our 'shape'. Over these we have a certain degree of control. An obvious example is the food we eat. The slogan 'we are what we eat' must be very largely true. But it's not the whole truth. 'We are what we do or how we conduct ourselves physically' is true as well. As adults we choose whether we sustain our bodies in good shape or trash them.

Our baby bodies 'know' when it's time to go exploring and when crawling isn't adequate for the purpose. Babies recover quickly from colds and fully from broken bones – they are remarkably mobile and resilient. As children, it never crosses our minds that our bodies might stop being healthy and efficient. Assist an aged person to get in and out of bed, however, and you begin to wonder how the innumerable muscles involved – from toes, feet, ankles, knees and so on through the whole body – could ever have been instantly and easily mobilised without due care and attention. No amount of neural signalling or biochemical alchemy will restore the postural workings of a body that has deteriorated. Will-power might make ageing wheels grind on a little longer, but only by forceful means. We use the terms 'will' and 'will-power' to denote determination or *conscious effort* as if, in recognising that we can't just get on with something in the normal way, we need to summon help. What keeps us going in the *normal* way?

So long as everyday activity remains spontaneous and unproblematic, we have reason to believe that our bodies are healthy. While we're young we tend to take health for granted, but it's during the early stages of growing up that various aspects of life negatively (if imperceptibly) affect the efficiency of our bodies, challenging Nature by requiring her to operate under unnatural conditions. An obvious example is sitting. We're designed to sit but most chairs are poorly designed for sitting! I call examples like this 'negative civilisation'. While individually we may not be entirely responsible for a life-style which is anti-Nature, we can be made aware of misuse of our body and its consequences, and find ways to counter it and repair damage.

I have already talked about the deleterious effect on the voice of learning to speak (in Chapter 2) and the physical and vocal inhibitions brought on by emotional repression (Chapter 3). The body (and therefore the voice) loses its shape through neglect and misuse. These issues – intellectual, psychological and physical – may feel burdensome enough individually, but they are also mutually affecting. While psychological trauma or distress can bend us out of shape, poor posture can influence our state of mind. As speaking doesn't recognise the voice as a whole, the vocal musculature of our throats weakens and seriously affects our breathing, which in turn affects our emotional freedom. It's impossible for vocal difficulties to be related to a single area. In our field, as in others concerned with health, one area of concern can illuminate another, broadening diagnostic terms of reference, and helping us to reach the roots of problems.

The human spirit

There are those who believe that a negative attitude produces negative consequences and that positive thinking leads to positive outcomes; *how* we think determines *what* we make of our lives. Whatever the individual case, there seems to be a natural impulse or 'desire' to live, which in every one of us is more or less up and running. We could therefore say that we stand upright because we *desire* to do so. I'm not talking about some kind of evolutionary expediency with regard to the senses, communication, and all things manual (which no doubt can be proven), but about the *human spirit*: that spirit which, in spite of the dangers to which standing on two feet exposes us and any emotional 'baggage' it may be our misfortune to inherit or pick up en route makes us defy gravity, hold our heads high and stride forth to meet life. This is the human attribute that, for example, sustains the dignity of people in poverty, misfortune or old age, inspires scientific investigation, and motivates physically handicapped athletes.

Historically, science and metaphysics have parted company at this point but it is where, perhaps, they have most need of each other. While physical vulnerability and dis-ease might seem easy to explain and the human spirit might be thought to be indomitable, rarely do we consider what effect the one might have on the other. It is one thing to survive the 'slings and arrows' and quite another to make the most of what we have. There's never a greater chance to do this than when we're young. However, considering that singing – if we want to fulfil our potential – demands so much of us, it's likely that we'll need time as well as determination to achieve our goal.

When we think about posture it's usually in mechanistic terms. A definition of good posture might therefore be 'the healthy employment of the mechanical body or the mechanical deployment of the healthy body'. Merely talking about it suggests that not only is it somehow important, but we can do something to *affect* it. As practitioners concerned with the health of bodies, it's important to note that when some activity gets problematic or we feel weak, we are generally inclined to use will-power or even force to *overcome* the difficulty. Ironically this is aided, if not abetted, by our body's versatility and adaptability; in response to our wilfulness it finds a way of compensating for physical weakness or ineptitude.

Hand in hand with our mechanistic approach to the body goes our thinking in terms of the *effort* rather than the ease involved in physical activity. Ironically, we give no more credit or credence to the body's natural ability or grace than we do to the neuro-physiological auto-pilot which turns our intentions into physical actions. We prefer to control our bodies rather than let them work, suggesting a lack of balance in favour of mind and will over matter and spirit.

It could reasonably be said that the degree to which a person takes care of her or his body is the degree to which she or he respects it, or even life itself. Why do we pay so little heed to our body's needs when we depend upon *it* for so much? We seem to create a vicious circle: we neglect our body and so it doesn't work as well as it should, therefore we cannot trust it so we push it or force it to work until it breaks down. Thus we prove that it's untrustworthy. Why do we maltreat the very thing that our will-to-live depends on to do the living? Modern life, which promises to provide us with everything we need including cures or solutions for things that go wrong, is largely to blame. A consumerist, mechanistic mentality supports this view. Physically it seems hardly necessary to lift a finger. We don't even have to walk very much any more. If, however, we really want to do some activity well, we must understand that we're not machines but something organic. In attending to posture we are recognising the interrelatedness and interdependence of the various parts of our organic selves, and acknowledging that no one else can do this for us.

Here are some of the reasons, relevant to our present context, why we don't give our amazing bodies full credit for what they are able to do by nature:

1. **The work ethic:** working for the sake of working, leading to the idea of perfection, and misplaced effort.
2. **Curiosity about *how it works*:** understanding intellectually *how* something works can cause us to interfere with natural processes.
3. **Control over nature:** the perceived need to control and dominate leading to blocking or forcing action, undervaluing spontaneity.
4. **Denial of the physical body:** our attitude to our body and sexuality can adversely affect necessary physical involvement.
5. **Consumerism and competitiveness:** wanting to 'be best' or 'get the best' can create false values and expectations, leading to end-gaining and superficial results.
6. **Employment of unnatural means:** not understanding the *nature* of what we are dealing with can lead to irrelevant, pre-planned work strategies.
7. **Lack of education:** lack of knowledge about and appreciation of our bodies and their needs.

It may help to understand that what sustains us in life is movement. Even when we are asleep the body is pulsing away in a perpetual state of aliveness. It is constantly reorganising itself by means of waves of 'information' in response to external and internal stimuli and most of this activity is unconscious. We can assume that this is how it is *meant* to be. Even when the day begins and we become more obviously active, performing one complex activity after another, all movement is taken care of. Until, that is, we become aware that our bodies are experiencing difficulties and we begin to interfere. As our bodies age, to give one example, we may be obliged to take more note of *how* we get out of bed or into a car. Until this point, unless we have consciously kept fit for some specific purpose, there's a gradual, if imperceptible, decline in our physical health and efficiency. Why shouldn't life keep us in agile, economical, graceful shape?

From the moment we begin to stand up, run around and our fingers fiddle with everything, it is clear that evolving fluidity and dexterity is part of our design. Why can't we assume that our brain will keep us this way? Unfortunately, life's many influences do not all complement the promising start that Nature has bestowed. These influences, which can be as diverse as the twisting of an ankle or the death of a friend, are largely unpredictable. A lack of trust in our bodies is often linked to a more general lack of trust in the processes of life itself. Instead of being encouraged to enjoy life's intriguing and exciting journey, we devise means of protecting ourselves *against* it.

The conditioning of our spirit as well as its condition moment-by-moment shows in the way we hold ourselves, push ourselves or let ourselves go. Attacked, our spirit may react aggressively or defensively, bracing the body in fearful or defiant rigidity; humiliated it may bow the body in defensive withdrawal. Strongly or persistently offended we literally *embody* our reactions so that over time we assume their physical features. The insult, being literally held in tension, blocks our freedom to breathe and to express ourselves freely in motion. To use a famous quote, 'our biography becomes our biology'.[4.1] An example is our reaction to fear, which causes the body to cringe. If the fear continues the cringing becomes compulsive. The entire mis-shaping process happens unconsciously. In our concern with balanced, efficient, economical movement, we must address not only the causes but the negative consequences of the unintended shapes our bodies assume.

If a would-be singer doesn't rid herself of unwanted muscular tension she will compound it through force of will by compensating for the limitations in movement that it imposes on her voice. As teachers we should understand that in reconditioning a tense body we are removing a form of psychological or emotional protection.

Liberating the voice liberates the spirit as well as the body, and it may be that this process will inspire and enable the singer to rise above the original psychological cause. However, a singer's condition might indicate that some alternative physical or psycho-physical treatment is required. Whatever the work needed, premature muscular release can be extremely disturbing and inappropriate delving into a pupil's psyche may cause further retreat or instigate complications. However,

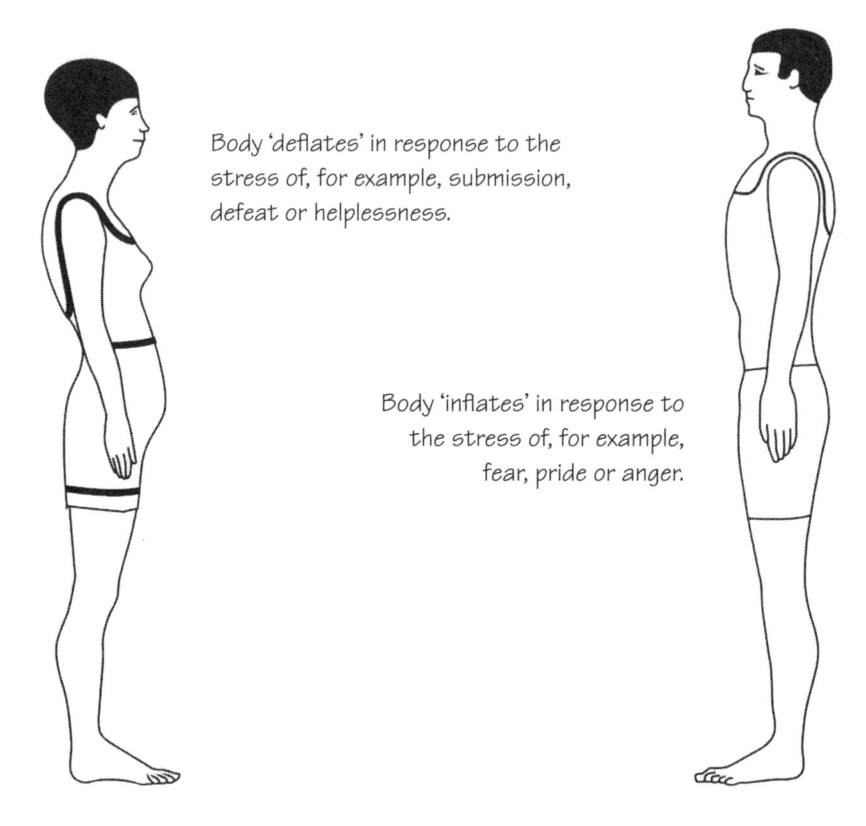

without going to the root of whatever is blocking a singer's singing, any treatment we are able to offer is bound to be limited if not merely cosmetic. The way people hold themselves often seems so much bound up with their psychology that even the mention of posture can put a person's 'back up'. Someone who has unwittingly adopted a defiant stance might 'dig his heels in'; another, bowed with shame, may be ashamed of his shame, and so on.

It is therefore beneficial to help pupils to understand the effect of any misshapenness on their singing, so that the measures taken are directed towards what they want as distinct from what they are afraid of. In any case, singing is so all-involving that tackling some problems while ignoring others makes no sense.

PART I

Chapter 5

The Inspiration of Life

Experiment 5. 1

Take a deep breath.........and.........hold it!

Note the pressure of breath in your thorax.

Now, keeping both your mouth and nasal passages firmly closed, try breathing out.

What do you feel – apart from the impossibility? The inward pull at the level of your lower ribs and abdomen gives rise to the sensation that pressure is being forcibly directed upwards and downwards simultaneously. Observe that all the movements gel automatically as one coordinated physical gesture. Has this a familiar feel to it?

'Potty training'

You cannot make a young child use a potty. However, when his body decides to 'go', no instructions are needed. Defecation happens quite naturally. To suggest that a child take a deep breath, close his ventricle bands and bear down might, if he understood a word of it, produce confusion, or make him laugh. Being oblivious of the need to practice any civilising process at all, he's just as likely to 'sing' or get up and toddle about. Laughing, crying or singing, he'll not pause to fill his lungs with breath, since all these activities also come quite naturally to him. He can sing and cry without vocal fatigue or damage, far longer than his Mum!

Defecating is not the only activity in which there is strong coordination between the body and the throat: coughing (to forcibly eject foreign bodies from the trachea) and stretching movements, for example, call for an increase in thoracic pressure. This pressure of breath is resisted at the larynx not by the up-turned vocal cords but by the down-turned ventricle bands, aptly called false cards.

Clearly the true vocal cords, with their upturned margins, were designed to close against the entrance of foreign bodies, including water, into the lungs. What's less well known is that they close whenever the breath pressure in the thorax needs to be *low*, preventing air entering the lungs. If you've ever tried lifting anything at all heavy, you will have experienced the physical logic to this state of lowered reath pressure. Try lifting having first increased the pressure of breath in your thorax and you'll be lucky not to rupture something!

We can equate the strong closing of the true vocal cords with those circumstances in which the muscles of the chest and arms need to make the most of their strength. The physiology involved in these activities ensures that oxygen continues to be fed to the blood stream. But how often do we attempt heavy lifting or use our arms for climbing or strong throwing? Most of us are inclined to avoid physically strenuous activities altogether.

The structure and primary purpose of the larynx was and still is that of a two-way valve. The original rudimentary structure, in place before our amphibian ancestors emerged from the oceans, went on evolving. That evolution was clearly for something other than bodily economy related to expiration and inspiration. Remove the need to kill our dinner with a spear or communicate over a great distance, consider the fact that our developing minds commandeered this highly evolved structure for its own specific purpose, add the abnegation of vocal expressions of emotion and an increasingly sedentary life-style, and we have plenty of substantial explanations for the gradual muscular weakening, over millennia, of the whole vocal tract.

A vestige of our original physical strength can be witnessed in babies, who are able to hold themselves up by the strength of their arms. Like their primeval ability to swim under water, this monkey-like activity is short-lived. However, a baby's remarkable ability to scream for long periods continues right through infancy. We say a baby has 'a good pair of lungs'. It is more appropriate to contemplate the work done by his or her tiny vocal cords: with these same cords that child might one day sing Siegfried or the Queen of the Night.

Experiment 5.2

Instead of taking a deep breath, exhale the air from your lungs through your mouth. This will induce a muscular response from your throat (perhaps a yawn) but shouldn't produce a death rattle or gagging effect. Accomplished *without resistance* at your throat, your entire rib cage will happily participate *without manipulation*. It may help to imagine that the breath is emptying from the top of your sternum, rather than through your mouth (which in any case should remain hanging loosely open). If, like most people, you are not used to emptying your lungs, go as far as you can without strain. If you find it easier, breathe out through your nose. This exhalation will take between 2 and 6 seconds.

Why not start again! Exhale – this time through your mouth – to your greatest comfortable limit without force or throat-tightening. Let your cords close. Then, in the middle range of your voice – without taking a breath – make a short, sharp, unaspirated, open 'Ah!' as though in surprise.

This primitive vocal gesture might be described as an exclamation or grunt (similar to the sound sometimes heard issuing from the throats of tennis players!). In a somewhat more sophisticated version the Italian School called this a *colpo di petto* (literally 'stroke of the chest'). (*Caution:* This is neither an aspirated sound nor a glottal stop (*coup de glotte*) as such.) Correctly executed, this muscular activity causes no discomfort whatsoever; the 'stop' is sensed below the throat, at the top of the sternum or a little lower. It's not made by an explosive force of air – there's not enough pressure for that. It's an 'implosion' rather than an 'explosion'. DO NOT PRACTISE this – if it doesn't work after two or three attempts, persistence is likely to do more harm than good. The *colpo di petto* only becomes a useful exercise when the outer muscles of the larynx are well developed.

It's a remarkable fact that your vocal cords can produce a strong, commanding sound, effortlessly and comfortably, with practically no breath. This, we will see later, turns out to be of vital significance.

The experience of breathing out

Experiment 5.3

Take a deep breath again … hold it for a moment … let it all spill out at once like a big, unvoiced sigh. Take another deep breath, and, after holding it for a moment, with your lips very lightly pursed let your breath escape softly without holding or squeezing. (At the end you may or may not want to take another breath.) What you have done is essentially and deliberately:

1. Inhaled more breath than usual.
2. Merely let this surfeit of air escape, somewhat passively, and returned to a position in which you still have two lungs full (not over full) of air – to 'neutral'.

Experiment 5.4

Lie on your back as if about to go to sleep, relax and calm yourself. After a little time, and without interfering, observe your breathing. What happens? Little – but that little is very precious, and turns out to be highly significant in our study of breathing per se. Alternatively, observe a baby breathing.

Our natural, relaxed breathing pattern is **out-in, out-in,** the exhalation cycle being longer (if only slightly) than the inhalation cycle. I believe that this order was established from the moment we expelled the amniotic fluid from our baby lungs in exchange for our first breath. These out-in cycles need no assistance; on the contrary, if they didn't work of their own accord we would struggle to stay alive.

The significance of the fact that breathing **in** follows naturally from breathing **out** cannot be over-estimated since it indicates the only major problem that civilisation has with breathing per se. As the **out-in** pattern calls into question the whole concept of 'support', and in view of the larynx's primary valvular role, we are bound to examine the effect of laryngeal efficiency or inefficiency on breathing, as well as the true role of the diaphragm. We will discover that training is truly 'potty' that concentrates on inhalation and volumes of breath.

More breathing out

There are many breathing out muscles. Yet, to most of us, breathing out with any strength or to any appreciable extent is unfamiliar: we rarely lift heavy objects and the pain of side-splitting laughter is, alas, a rare luxury. I once heard that the average human being normally uses only about 15% of his exhaling capacity. If it was 50% we might think we had cause to worry, but 15% seems alarmingly close to extinction! The reasons for this generic weakening of the human's respiratory system may already have become clear.

Experiment 5.5

Without deliberately taking a breath, exhale through loosely pursed lips, with a gentle, windy 'fff', quite freely and without holding back, until you feel muscles contracting at the level of your lower ribs, back, sides and front. At this point, without impeding the smooth flow, change to 'sss' and continue, without force, until you feel completely emptied of breath, and back, sides and front have contracted to their full extent.

What happens now? You have reached the limit of exhalation, and have no choice but to let your body breathe in. In other words, you don't make a conscious decision to inhale – you have no alternative but asphyxiation! Furthermore, this reflex action has completely replenished your lungs with oxygen, and you are back to neutral without over-stretching them. The effect is similar to squeezing the air out of a dry sponge; on releasing the sponge it instantly resumes its original shape. Notice in passing, however, how much breath can be exhaled from your lungs even when you have not deliberately breathed in. **Caution:** If you're not used to breathing out so far this could cause dizziness or panic!

Experiment 5.6

Without preparing, blow suddenly and strongly, with a short puff as though blowing out a candle. Again without preparing the breath, make short puffs (separately and in quick succession) as if blowing out several candles in a row. Once more, with shorter and sharper puffs – but don't rush it! Beginning without preparation, with a relaxed and energised body, is crucial to this experiment's success.

The most important thing to notice is that it is unnecessary to take breath between these exhalations – it's taken for you, and in theory you could carry on for ever without deliberately taking another breath, since each outward puff triggers the in-breathing reflex.

When attempting this experiment for the first time, most people struggle with it, or at best find the movement sluggish, or else they try too hard, and cannot relax between each puff. This is hardly surprising – it demands the rapid coordination of a great many muscles. Although the 'puff' is a natural respiratory gesture, it has become weak through insufficient stimulation; we're not used to breathing out from 'neutral' with any strength.

Breathing in

Inhaling requires no conscious effort providing the breathing out system is sufficiently well innervated. At the farthest reach of exhalation the diaphragm (our primary breathing-*in* muscle) is simply bursting to take over. The pain experienced in deep laughter comes as much from the denial of this reflex action of the diaphragm as from the exhaling muscles that are working so hard at their limits. It is hardly surprising that one feels that one might 'die of laughing', not so much from the pain as from the prevention of the diaphragm from keeping us alive!

Singing and breathing

It is commonly assumed that there's a special way to breathe for singing, which entails increasing the capacity of our lungs. This might seem logical, particularly if (as is normally the case, especially with inexperienced singers) we don't feel as though we have sufficient breath to sustain a phrase of music, or to sing loudly. However, if we examine the breathing system as a whole, we can begin to understand that this is a fallacy which has led to much abuse.

In normal everyday life the most obvious role of breathing (supplying our bloodstream with oxygen) is barely considered – we take this vital work for granted. It's only when we find ourselves unusually short of oxygen that we become conscious of our breathing and the physical effort involved.

All physical actions involve our breathing apparatus. We have already considered examples of strong muscular interaction in cases where the breath pressure in the thorax needs to be high, and the less obvious, less practised case in which the pressure needs to be low. In the former, we may be more conscious of the role of the breath, in the latter we are more likely to be conscious of the muscular activity that it facilitates.

Experiment 5.7

The next time you are puffed running to catch the bus or hastening up steps, try exhaling (rhythmically blowing) and allowing the air to be drawn in as a natural consequence. Remember, it's the oxygen we need; a great volume of air will only impede the operation of our limbs.

It should be clear by now that the *efficiency* of breathing depends upon the larynx. From this it can be understood that if, when attempting to sing, we imagine that it is a question of breath pressure, we will trigger an automatic response from the ventricle bands, aptly called 'false cords'. In other words we instigate a laryngeal dilemma. With increased intake of breath, the false cords have a job to take a back seat. Unlike in yawning, when they're deliberately drawn apart, they are being asked to participate in an activity for which they're not equipped. In order to make them comply, we 'devise' further ways of holding back or 'reserving' the extra breath, while taking pains to prevent the ventricle bands closing.

Along with such familiar terms as 'open throat' and 'support', both 'rib-reserve' and 'sub-glottic pressure' have become accepted parlance among voice teachers and voice scientists. These terms describe practices which (alas!) easily find themselves at odds with one another; banded together they can and do cause endless respiratory confusion and struggle. The term *inhalare la voce* (literally 'breathe in the voice') coined by the Italian School, from which the idea of rib-reserve may have derived, was on the other hand an imaginative way of balancing the *muscular dynamics* of breathing. It had, I believe, nothing to do with the now favoured hoarding or 'measuring' of breath.

While the simple ventricle bands have, through regular practice, retained sufficient strength to do what they were designed to do, the same cannot be said for their relatively much more complex and sophisticated downstairs neighbours the vocal folds; not, that is, beyond their original role of physiological 'bouncers' at the entrance to the respiratory tract.

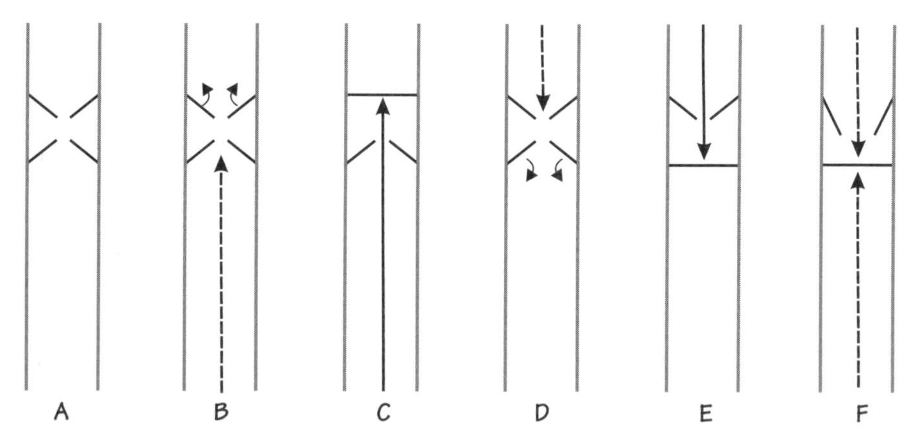

A: False cords positioned above vocal cords

B and **C:** False cords close against pressure of breath

D and **E:** Vocal cords close to prevent entrance of foreign bodies or to facilitate strength in upper body

F: Balance of outward and inward movements: the situation in singing

Coordination

In the rudimentary valvular work described above, the body and throat work together – they need each other if the task in hand is to be carried out efficiently. If the throat is weak the body's task is made less easy, and vice versa. This mutual dependency is of importance in understanding the muscular dynamics of the singing voice, and in being able to allot and apportion restorative work.

It is instructive that the throat-body coordination of the 'high thoracic pressure scenario' has been maintained throughout years of almost daily use – the natural part of potty training, after all, needed no encouragement. What of its opposite, the far less familiar 'low pressure scenario'? Wasn't this also something we were born with? In a baby, this primitive strength is demonstrated through the infant's capacity both to hold itself up by its hands, and to vocalise for long stretches. (The former capacity rapidly declines, even though the gripping reflex remains.)

Why we have to coerce our voice into action for singing, and what has taken place to reduce the natural strength and limit the functional response of the human's singing cords has been the main thrust of this book so far. If we know why voices in general are weak or inefficient we may be able to see more clearly how to restore them to their natural healthy working order.

Breathing and singing

Experiment 5.8

Take any sentence from the text you're going to sing and simply say it.

Did you have a problem with breathing? Of course not! It was in fact little different from conversation, when you are rarely conscious of taking a breath, let alone a big one. Even in a crowded pub you don't need significantly more breath in order to be heard above the din. In normal everyday speaking, however, more air 'escapes' between the vocal folds (which are relatively slack or ill-tuned) than in good singing, when they are firmly aligned and finely tuned. Is there not a contradiction here? In speaking we scarcely breathe, and yet we 'use' a lot of breath. In singing, less air escapes, but we seem to need more.

The answer appears to relate to the fact that in singing more is going on. Is singing just more demanding? (You have to vocalise louder, or higher, projecting your voice much further than across a table.) Does this 'more' really depend on the *quantity* (and perhaps the force) of breath used? Infants certainly do not need more breath than their little lungs already contain. One might argue that most of their 'phrases' are short, but even so, how do their voices carry so piercingly far on so little breath?

Earlier we experimented with the making of a vocal sound with practically *no* air in the lungs. Try it again:

Experiment 5.2 revisited

Breathe right out and let the cords close silently. Then, without an aspirate, exclaim a short sharp 'Ah!' as though you are ecstatically surprised to see somebody. What a strong sound you can make with no breath!

What then *is* the difference in the amount of breath required between speaking and singing? When we stock up on breath we generally do one of two things:

1. 'Press' the voice with it.
2. Try to 'hold on' to it.

Both these measures are crude attempts to 'take charge' of the amount of breath we use. Hoarded or 'reserved' breath and breath put under pressure demand conscious muscular effort and, as a result often provides a *feeling of support.* In both cases, however, freedom of movement between the body and the throat is impeded, neither facilitating the coordination between the throat and body, which should be sharing a mutual purpose. These ways of manipulating breathing may produce a sound which we call singing, but since both methods are potentially debilitating, it's hard to say which is the more injurious. The practice of taking more breath than the singing voice requires weakens the tonicity or natural elasticity of the lungs. Pressure and 'holding' cause unwanted tension in the throat, which is then hard-pressed to provide the freedom of movement upon which the larynx thrives.

If what enables us to sing without force or unwanted tension is not the breath, either held or pumped, it can only be the action of muscles, *regardless* of the amount of breath used.

There was a time (perhaps as little as 20 years ago) when some teachers abhorred the idea that singing was anything to do with muscles. Perhaps they represent a stage between those teachers who possessed a sure aural knowledge and had no need to think in terms of muscles, and those who have begun to depend increasingly on scientific ideas. Few teachers nowadays would doubt that in singing we employ a great many muscles. However, we should beware of the pitfalls in viewing such a vast complex as the human voice in this way. 'Muscling in' to the voice is a symptom of misunderstanding the purpose of its muscular structure, which among other things is there to make vocalising easy. Fixed or forced vocalisation results in only a crude semblance of singing. Strenous forms of muscularity, in demonstrating the conflict between the body and the throat, cause a voice eventually to seize up or break down.

Popular jargon phrases such as 'columns of air' and 'sub-glottic pressure' are not user-friendly. When you consider the *smallness* (both in length and bulk) of the average vocal fold system, compared with the rest of the musculature involved, you can see that it would take little effort to put the folds under undue pressure, or even 'brutalise' them. The *struggling* that often seems to be encouraged (if unwittingly) in singing, suggests that the relationship between breathing and these short lengths of muscles is not well understood.

In repose, the cords lie passively open, allowing air to pass freely out and in (precisely what occurs in simple panting). If you close the cords, or rather allow them to close (as in Experiment 5.2 above, or in a proper, un-explosive *coup de glotte*) you realise at once that the amount of air able to pass through them is determined simply by whether or not they are closed.

The popular perception is that pressure of breath is responsible for both adduction and vibration of the cords. If this is true, it must be acknowledged that the only way in which the relationship between the voice and the body can be *effortlessly* achieved (without undue tension and force) is if the adduction is efficient and strong, and the 'resistance' that vibration implies is minimised. The active opening (as for example in yawning) and closing (as in the *coup de glotte)* of the cords can be clearly observed as muscular movements *unaided* by breath.

Unfortunately, normal vocal folds are not in the habit of adducting with optimum efficiency, let alone strongly. In *singing*, the body gets the message to 'help out' when the cords *lose* their adducting strength. This appears to be a physiological law and the body kicks in to secure a firmer relation between itself and the throat. In any case, if, when we try to sing, the *response* from the voice is weak, it would seem natural to try to use a greater quantity of breath, in order to press our cords into action. This would then pose the questions not whether the extra breath is acceptable so much as 'how much breath?' and 'how do we control it?'

Breath control is talked about as though the breathing apparatus must be manipulated. This is why 'special breathing for singing' has generally become of great importance in the eyes of singers and teachers. Unlike our vocal folds, breathing is by and large something that we can both feel and see. If a weak larynx 'calls for support' from the body – from the breathing department – it seems logical that the more breath we have the better, especially to support the voice through phrase after long phrase. It seems logical too that the stronger the engagement between the breath and the larynx, the louder and higher we will be able to sing.

However, the body is receiving mixed messages and its impulses are confused. We've forgotten the 'high thoracic pressure scenario' with its strong response from the false cords. We have overlooked how weak and inefficient our breathing system has become, in particular the 'low thoracic pressure scenario' with its natural response from the true (vocal) cords and we may have ignored the feeble or inefficient support that the larynx normally receives at its own level. The deliberate stocking up of air seems to be an unconscious attempt, sometimes a desperate one, to make up for widespread systemic failure.

Those whose instincts warn them against this idea invent various techniques intended to 'hold breath back'. Instinctive or not, this idea gives the larynx more freedom of movement. Because of the *inhaling* (open-throated) nature of such measures, they are sometimes partially successful, especially in persuading the false cords to get out of the way. If the free vibration of the vocal cords is the main or only aim, 'holding back' methods can go some distance towards achieving the desired end. However, singing being so much more than a sound, we cannot afford to hold our breath – or indeed anything else.

Scale

Have you any idea of the size of your vocal cords? Most singers I've asked have not considered this question, or even realised that the cords lie horizontally in the throat. There's no need for singers to dwell on these things, but if we are deliberately working with these muscles, their size might influence how we treat them. Estimates vary slightly, but one is that the maximum length of cords in a male singer is 23mm (———|——) and in a female 17mm (———|——).[5.1] These lengths are divided into 3/5ths and 2/5ths since it is the membranous 3/5ths only which vibrate. Logically, the longer the cords, the lower the voice: sopranos have shorter cords than mezzos, tenors shorter cords than basses. (The capacities and qualities of a voice depend also on other fold measurements.)

Not only are vocal cords small, they're remarkably complex. Robert Sataloff, in his comprehensive work *Professional Voice,* states 'the more one studies the vocal fold, the more one appreciates the beauty of its engineering'.[5.2] In addition to being responsible for the sound we make and the pitch at which we most comfortably make it, these short lengths must contain, within their narrow limits of flexion and extension, all the notes of the individual's natural range, usually somewhere between two and three octaves. They must make adjustments between many intervals at both fast and slow tempi, whilst maintaining the sung tone at constantly varying dynamics with a rainbow palette of tonal colours. As if all this were not enough, they must reflect human emotion in all its depth and subtlety.

The idea that all this can be achieved satisfactorily, let alone easily, under pressure or by any fixed or strongly resistant means is inconceivable. The vast supporting structure employed in the act of singing – extrinsic muscles of the larynx, specific breathing muscles and postural muscles – is designed to *facilitate* the fine and complex interplay of events in the folds themselves – to *reduce the effort.*

To obviate struggle, it must be understood that the relation between breathing and the throat, and by extension between the body and the larynx, is a natural one. If it is found that the breathing system is weak, shallow or lazy, or if something in the throat or larynx is flaccid or stiff, these issues must be addressed. The co-operation between the various spheres of activity – larynx, throat and body – must be continually tried and tested if one of them is not to be more dominant or dependent than another. This work will be explored in Part II.

As the work proceeds, it is realised that the amount of breath passing through the tiny pair of cords is so minute that it barely seems necessary at all, and that singing 'with breath' or 'breath support' becomes a redundant pursuit. For singers whose voices don't seem to respond in the way they would like, and for those in the *habit* of employing a lot of breath, this small amount seems far from their experience and scarcely plausible. Depending on less breath may require an act of faith, until it becomes second nature.

With assiduous training the 'long tone' much prized by the old *bel canto* school (as distinct from the 'long breath' sought after by modern trainers) feels gradually attainable, as little by little the idea of being 'at one' with our voice materialises. Breath can be perceived more and more as a carrying medium rather than as a motive force. When singers reach this experience, they begin to relax about such things as 'projection' of voice, volume, and even those hitherto unattainable 'high notes', none of which can satisfactorily be achieved by increasing the volume of breath. Even if it's not yet understood how to arrive at this scenario, the danger in giving undue importance to the amount of breath we need in singing should by now be clear.

The idea of 'singing on the breath' (as distinct from *with* the breath) possibly arose from the ease with which a voice in the 'minimal breath' condition flows, with its accompanying lack of pressure and absence of 'fixing tension'. The implication is that the voice flows as the breath does, and while this may sometimes prove a helpful image, encouraging a more relaxed 'delivery', it does not necessarily bring about efficient vocalisation. A singer remains short of the mark so long as there is a real sensation that air is flowing. When the breathing apparatus is used wrongly we get the *impression* that breath is all-important, but when it plays its natural part without bidding, there is no such impression. The amount of air is infinitesimally small, as befits its delicate and highly specific task at the level of those tiny cords.

In singing it's only when breath is evident, either in being dammed up or because it is 'leaking', that we should be concerned. To attempt to control the breath without the natural participation of the vocal folds is a folly, and the cause of much misplaced hard work, and damning struggle.

It would be convenient to be able to say that if breathing per se is 'correct' the voice will respond accordingly. Unfortunately, to function correctly in singing the breathing system needs the larynx and its extrinsic support system as much as the larynx needs the breathing system. 'Breathing athletes' with exemplary posture (such as yogis) may have their breathing very much under control but cannot sing.

While breathing per se (which humans usually do so badly) is a matter of life or death, it seems that the singing voice, which is such an efficient promoter of good breathing, is not. How perverse and frustrating, that while the *desire* to sing prevails, this system has become so weak! An intriguing contradiction!

PART I

Chapter 6

Unavoidable Conclusions

Introduction

In teaching singing I have realised that people have many reasons for wanting to sing. 'I want to learn to sing' can mean 'I want to sing better because I enjoy it so much', or 'I want to be able to sing all that gorgeous vocal repertoire', or 'being a singer is glamorous – I want to be a star!' It might also mean 'I want to express myself' or even 'I want to find myself'.

My concern as a teacher is with the work involved in reaching the highest and deepest levels of vocal freedom and expression. While even professional singers don't always reach a high degree of vocal freedom, I have to assume that my pupils want to sing to the best of their capabilities, and find out what this means for *each individual.* Each singer embarks upon a unique journey, which not only paves the way for music making and communication through the voice, but invariably amounts to a therapeutic or healing process. Releasing her or his singing voice releases the singer.

While the idea of human progress for its own sake appeals to me, as somebody who loves music and singing my chief motivation is the desire to experience great vocal music at its best: music performed with skill and accuracy, freedom, depth and joy, artistry and attention to detail, and with passion. These are only some of the qualities of which truly committed, effective communication is comprised. I don't want simply to be entertained, but to be *moved*!

One fundamental reason why learning to sing can seem so problematic is that it is personally revealing. Singing reflects our condition – how together or untogether we are and in what ways – as a human being. In the first part of this book I have described the various ways in which voices have suffered because of civilising or cultural influences and suggested that they betray in some respects the degeneration of the human species. Voices get split into various, more or less separate, parts, demonstrating in sound how we are divided, and how we suffer from being so.

In liberating the voice by enabling it to be itself, we have to deal with adverse conditioning of our minds, bodies and emotions. We must confront our inner selves and recognise our shared humanity. This is unavoidable if training is to be thorough; learning to sing is bound to be therapeutic for anyone taking it seriously.

Much has been written about singing as therapy, and there are practical courses exploring the therapeutic properties of vocalising. Although many books have also been written about singing, the weighty task of what is really involved in becoming a professional classical singer has not yet to my knowledge been adequately addressed. Would-be singers rarely have any idea of the demands of the profession, let alone the nature of the training that they must undergo to have a truly satisfying, let alone long, career.

For most people, learning to become a good singer is hard work because it involves so many skills: musical, vocal, linguistic, dramatic and so on. Work on the so-called 'technical' front alone can seem daunting. Although there may be no holding back someone who passionately desires to sing, a lot of disillusionment and pain can be avoided if a student is given sound advice from the outset of training about what he might reasonably expect of himself and his guide. The journey is largely unpredictable, and challenging. Lack of advice leads to waste of time, money and energy. Inspiration, determination and enthusiasm are merely pre-requisites in learning to sing. They fuel the process but they're not *it*. Only when singers and teachers have taken the trouble to prepare mentally and physically for their work can they expect to be satisfactorily rewarded.

Teachers and singers alike tend to underestimate both the challenges of learning to sing and the possibilities of the singing voice. From the gloomy picture I have painted of the normal voice's dislocated condition, as well as being able to gauge how far short of their natural state voices have generally strayed, we can imagine what the healthy whole might be like, from its basic shape to its full splendour and fine detail, as in archaeology when fragments of a fresco or shards of a pot suggest the shape and colour of the whole beautiful artefact. We also know that, barring irreparable physical impediment, the whole singing voice is potentially still there and retrievable. While it cannot be reconstructed artificially, it can be regenerated naturally.

If we can also see by what processes vocal deterioration has come about, we already have a notion of what may be required in restoring a voice to its singing condition. We have a helpful, and often quite detailed, perspective of each voice we encounter, with areas or points of reference against which we can measure its shape – the individual version of the norm.

Natural, naturally

Until recently, laryngologists peered down throats, pronounced vocal cords perfectly healthy and wondered why they couldn't sing. Speech therapists, having skilfully restored voices to a healthy condition, might suppose that something extra

– an *extension* of the normal – is called for in order to sing. The medical world, having determined what constitutes vocal normality, concerns itself with vocal pathology, while other researchers into voice often rely heavily on a comparison between this medical definition of 'normal' and what is deemed to be '*ab*normal'. The consensus seems to be that the singing voice is somehow *supra*-normal.

Is what is generally considered normal necessarily natural? Can speech and singing both be natural? If the answer is 'yes' we must ask why so many people with seemingly normal throats with regard to speaking cannot sing. In the 1990s, after a talk I gave for the British Voice Association, I asked how many of the audience thought that the singing voice was natural. Out of about 70 (mainly singing teachers and speech therapists) no more than five hands went up.

There's plenty of evidence to support the 'not natural' lobby. The implications for voting 'natural' may be frighteningly far reaching. If singing is natural, we must ask the question 'why are so many voices not working according to design?' Furthermore, accepting the singing voice as natural might create a dilemma for those who have trained to restore pathologically damaged voices to a *normal* state: they would have to consider that what has hitherto been recognised as vocally normal is not in fact as healthy as they supposed.

It's clear that between the normal, non-singing throat and that of the natural singer there are many different states of functional efficiency or inefficiency. Some voices are near to being 'natural', while others are nearer to 'normal'. This suggests that we should be looking for degrees of vocal health and effectiveness, which would put the true singing voice (the voice we are born with) at the top of the scale and the normal, speaking, non-singing condition (not yet described as pathological) somewhere near the bottom.

If singing *is* natural, we are still faced with the question of why so few people can sing. I do not believe that a firm, well proven answer will emerge from studies of vocal pathology – it's looking too hard at the wrong side of the 'health coin'. Nor will cohesive evidence be gleaned from comparing the differences between the speaking and singing voices of a normal non-singer. Surely it must be the operational *completeness* of a voice, as distinct from the absence of pathological symptoms, which provides us with the most comprehensive information about vocal health. Current voice research lacks sufficient available natural vocalisation, and the means to test it holistically. Until non-invasive, non-inhibiting investigations acknowledging the roles of emotions and the human ear are carried out on enough natural voices, answers with a practical use to singing teachers are more likely to be gained through the findings of psycho-social or psycho-physical research.

What can we consider to be natural singing? It's not a question of style, since styles are human inventions. It cannot be a matter of culture, since what is bestowed by nature must apply to all humans equally. Cultural differences in the

The training of non 'western' voices in the art of 'bel canto'.

use of the voice are stylistic, arising largely out of the languages and musical history of different societies. Can natural mean merely uninhibited? A non-singing voice must in some sense be an inhibited one. At my school, when we listened regularly to *Top of the Pops*, one student always joined in – loudly! His lack of inhibition made for an exhibition which (alas for his audience) was in every respect tuneless!

Western classical singing is often taken as a model for natural, healthy singing. We could describe it as an attempt to fully realise our vocal potential. The fact that classical singing is itself a style born of cultures could be a stumbling block for some, especially when so many examples sound far from natural. However, the 'completeness' (in terms both of skill and range of expression) of the 'classical voice' gives us clues about its universal as distinct from cultural authenticity, and its human as distinct from ethnic origins.

Non-western throats are just as capable as western ones of realising their innate vocal potential in classical singing given suitable liberating training: in recent years there have been notable examples among Chinese, Japanese and Korean singers, all of whose traditional manner of using the voice could hardly be further removed from, for example, *bel canto*.

Pop vocalisations are generally aberrant forms of vocalising. However extrovert the performers, their voices mostly have little adaptability. The classical counter-tenor is an example of 'incomplete voice': while he has a well developed *falsetto* his full voice is usually weak. A good tenor or baritone, on the other hand, using his whole voice, possesses a strong *falsetto* function.

It is worth noting that in the West the categorising of voices in classical genres (Early Music voices, voices for lieder, 'Wagnerian' voices, contemporary music voices and so on) has taken place in the past century or so, as singing has gradually lost sight of the vocal disciplines and aural knowledge gained in the previous three centuries. Alongside this decline, science has advanced sufficiently to explain and 'justify' new forms of vocalisation without having defined what the natural voice really is. The irony is that the proliferation of other forms reflects a cultural 'loosening up.' The resulting fragmentation demonstrates how easy it is to veer away from something centred, to reject the ability to *fully* voice ourselves, while also demonstrating the diversification of which the original voice is capable.

The rejection of the naturalness of the singing voice has led to different attempts to turn it into something technically contrived and alien to its true nature (sometimes in order to serve a more or less rigid musical or stylistic agenda). Most of these vocal 'techniques' replace the blocks or distortions from which a voice may be suffering with others that are more benign or seem more effective, based on a perception of the 'classical sound'. The ability to mimic a certain style however, classical or any other, does not in itself make for natural singing.

As singers and teachers of singing we must distinguish between musical styles and the physical emission of sound, and be clear about the differences (sometimes obvious, sometimes more subtle) between what is *physiological* – working in accordance with nature – and what is vocal but not according to the original blueprint. Our perspective must be such that we can clearly distinguish between the voice employed as a *natural entity* and vocalisation that employs the voice in part.

The natural voice can be defined by its melodious fluency, range of expressive qualities, physical skills and an honest whole-heartedness. None of this is attributable to training, except in so far as training means liberating what is innate. We instantly *recognise* the free voice, and on some deep level, *identify* with it. There is something quintessentially right about it, and we wish we could do *that.*

Breaking away from the so-called classical voice does not result in the freedom or range of expression that it might seem to promise. While vocal imbalances may be turned to impressive effect, unbalanced voices are noticeably limited in scope (musically, tonally and emotionally), and often require technological support. The fact that many of these sound-styles can be produced 'safely' and explained scientifically does not make them healthy – they are safe only in the sense that apparently they do no harm. Imbalance, however, is by definition unhealthy: it strains the whole vocal system and cannot be sustained without unphysiological compensation.

Compensating for incapacity

When we are incapacitated, our muscles and neuro-physiology often succeed in finding a way to cope. It's manifestly more difficult to compensate for some incapacities than for others. Try buttoning up your shirt or tying shoe laces without using your thumbs – not impossible, but tricky. How complex simple actions turn out to be!

The singing instrument is vast by comparison with the hand, but, were it not for the fact that it's a complex of muscle systems that serve many other purposes, we might think that it was far less useful. As the singing voice doesn't exist as an entity in itself, but instead depends on the coordination on impulse of many parts, we realise how vulnerable to imbalance it is. Without constant use, the singing voice as a whole lags behind several of its parts which take precedence in everyday life. For example, for various reasons, we 'practise' taking breath, then letting it explode through the larynx, or holding it and pressing with it (all decidedly non-singing relations between body and throat). We 'practise' holding on to our larynx and closing our windpipe many times a day in swallowing. At the same time muscles that should be used for vocalising are neglected, and we practise doing without them.

When we take singing seriously, we find there are plenty of muscles (in our throat or body) which are only too eager to stand in for those which have lost their strength through lack of use. The normal voice's inability to coordinate itself is exacerbated by these replacement tensions. This type of 'helping out' seems to happen initially on a more visceral than conscious level – the body feels the necessity and automatically summons its resources. Unfortunately when corrective measures become conscious or planned there's always the risk of compounding the problem, unless we acknowledge that these alien tensions are there for a reason. Trying to remove tension by focusing on it tends to produce further unhealthy or conflicting tension. If we recognise *why* tension is there, and *what* it is attempting to replace, we have a positive way forward: we will worry less about getting rid of it and more about what should be happening naturally to render it unnecessary.

When a voice sounds 'plummy' or 'back in the throat', we are hearing the result of substitute muscles acting for weakness in the suspensory mechanism. Instead of contributing to an elastic system which gives the larynx freedom to move, they attempt to hold onto it. This perversity may or may not give a feeling of support, but in fact prevents emotional fluidity and severely limits vocal flexibility generally. The tone produced is artificial and often sounds 'old'. Sometimes a variety of this sound is taught in the name of 'maturity'.

Muscles that compensate for a weak suspensory system comprise combinations of jaw-operating muscles, the tongue and other muscles normally employed in deglutition (some aptly called 'false elevators', since they lie above the larynx and tend to fix it from above). Little is gained by attempting to relax these. Instead, the true suspensory muscles must be re-invigorated and fully reinstated to make the 'false workers' redundant.

'Covering' the voice (in order to extend it upwards in range without 'breaking' at the so-called *passaggio*) also employs a fixing technique. Although at first such measures may seem expedient, as a way of bridging registers or making a voice 'rounder', they do not permit the singer to be vocally direct or 'forward', let alone tonally flexible or consistent.

In terms of the degeneration of the voice, the most severe division is between the throat and the body. One legacy of our 'civilising' is a weak suspensory mechanism – what in effect links the mind and heart, the spoken word with emotional feeling. Technically, this means that there is little to take the strain off the larynx when we want to vocalise strongly, and almost no 'support' of it when we wish to sing quietly. By being predisposed to pushing with the breath or fixing the system with extraneous tensions, certain ways of linking the body with the throat or the throat with the larynx can act like conditioned reflexes, and even give a sensation of 'support' for the voice. However, when in the attempt to sing the breath is put under pressure, the 'false cords' endeavour to close against the breath. While this action accords with nature, it is being called up for the wrong reason. Since we are *not* about to cough or empty our bowels, we struggle to counter this natural closing-of-the-airway response, being obliged to make sure that the airway *forward of the true cords* is sufficiently open. There could not be a clearer case of mixed messages. While strong throats can sometimes contend with this contradiction, the true suspensory system is prevented from doing its job with natural freedom. The result, audible to the listener and usually felt by the singer, is unhealthy stiffness in the throat.

When our bodies assiduously employ muscles in any activity they get stronger. Paraplegic athletes, for example, must propel themselves with their arms instead of their legs. In unbalanced vocalising we practise the weak as well as the strong. Weak muscles will get weaker, even to the point of atrophy, if neglected.

The deterioration of the singing voice in the process of acquiring speech is a good example of imbalance in which certain muscles lose their tonicity and become redundant while others, in compensating, become stronger. Those who treat the singing voice as an extension of the speaking voice, or something to be added, rather than seeing the speaking voice as a diminution of the singing voice, have set themselves an impossible task.

All aberrant voices are characterised by extremes of strength and weakness, by a lack of strong natural antagonism in the larynx, and by a lack of natural co-operation from the body. In unbalanced vocalising, the weak and the strong are practiced in tandem, perpetuating over-strength at one extreme and flaccidity at the other. This lack of balanced coordination is foreign to the true singing voice.

In our diagnostic work as teachers, the awareness and understanding of compensatory mechanisms and the imbalance that they are perpetuating is crucial. Since an unbalanced structure, assiduously practiced, can *feel* all right or safe, attempts at correction can feel wrong or threatening. Following the principle that such compensation has a cause, we have a reasonable chance of putting things right. Linking this understanding with knowledge of posture in general shows us that, although there are physical zones where compensation is common, the process of retrieving the singing voice is one of *all-round re-balancing,* from larynx to suspensory mechanism to body and back the other way.

Compensatory tensions don't confine themselves to areas of vocal malfunction. Lack of physical coordination necessitates misplaced effort when dealing with, for example, enunciation, dynamics or high or low notes, all of which should come easily to a fully liberated voice. Effortful singing – an overblown *forte,* diaphragmatic articulation, 'chewed' text, and so on – is by definition contrived singing because correct intentions have met with resistance instead of co-operation and only a degree of force or manipulation can satisfy them. All vocal contrivances betray themselves, either in discomfort or struggle on the part of the singer, or audibly or visually to the listener. A good performer might seduce us into thinking that all is well. However, we must understand that an uncoordinated voice is a limited instrument, and that no amount of compensation will prevent such a voice from deteriorating and eventually failing.

PART II

Sounding the Self

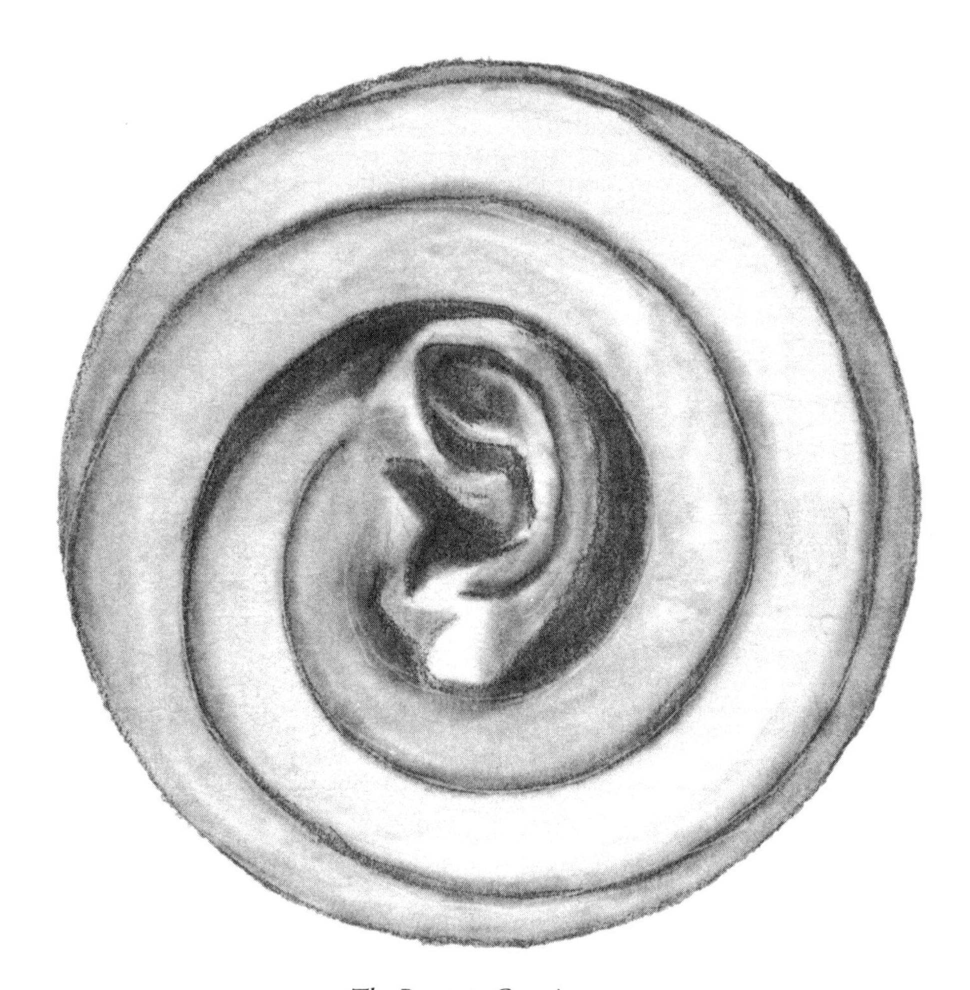

The Route to Consciousness

PART II

Chapter 7

Making Connections

Introduction

Part I points out aspects of the activities we call 'learning to sing' and 'teaching singing' which are easy to miss or underrate if we don't use our ears or open our minds enough. Important points to remember include:

1. Our singing voice is a complex instrument and by nature as vulnerable as it is vulnerable-making. Much can start to go wrong with it early in life, and restorative measures tend to be intricate and demand understanding and time.
2. The severity of vocal difficulties can lead to wrong conclusions, for example misguided notions about 'support', and the concept that 'registers' are part of the natural vocal order.
3. Current or habitual methods of work may not adequately address the full complexity or individual nature of the problems we confront. As teachers or pupils we must learn to detect in what respects a method is inadequate, weak or incomplete.
4. Not working deeply or broadly enough cheats everyone of what might be possible. Superficial work leads to unsatisfying, superficial results.

In the worst case we see a person's singing voice collapsed or deteriorated beyond recognition. At the other end of the vocal spectrum are so-called natural voices, which have survived the various negative influences described in Part I. In between these extremes are thousands of voices in varying states of decline or dysfunction. Many we will recognise as having a 'singing quality' – they catch our ear when they're a cut above the average, and some are judged special enough to merit training. At this point it may be too early to ascertain whether or not an individual has what it takes to become a professional singer.

In Part II I will examine more closely how the various spheres of vocal activity interrelate, how, for their own efficacy, they are interdependent, and how they *potentialise* one another to form something far superior to a collection of musical, vocal or performing assets. This I term 'the singing voice'. We will see that it is the voice's interdependent, self-potentialising nature that obviates conflicts, between (for example) vocalisation and emotion, or vocalisation and words, making instead for a wonderful creative synthesis.

Singing teachers talk about freeing or unlocking a voice, rightly acknowledging that (like the body that houses it) it is normally tense, uncoordinated or otherwise inhibited. Telling or even helping a pupil simply to relax, however, is rarely a satisfactory or lasting solution; tension banished from one area will usually set up camp somewhere else equally inappropriate. Tension from fear or hurt, for example, is likely to be a necessary defence or protection.

Interrelated spheres

If the various physical zones of the singing voice – breathing system, bodily alignment, suspensory mechanism and larynx – depend upon each other for their efficacy, we can infer that the singing voice, to be fully operational, depends upon the health and efficiency of these elements in their own right.

We have a postural frame, from or within which the muscles specific to breathing are slung, enabling them to work with natural ease, strongly and to their fullest extent, without the body twisting unnaturally or collapsing in on itself. Turn this around and we see that breathing, if executed to *its* full extent with sufficient vigour, requires the muscles of the postural frame to 'pull their weight'. Posture and breathing are mutually dependent: to achieve optimum efficiency and strength each must be given due attention.

Breathing *out* excites (or 'charges') the reflex of breathing *in*. The maximum efficiency of this vital system is achieved therefore by strong, extensive exhalation. Our 'desire' to breathe in logically increases as we breathe out. Expressing ourselves in sound is also primarily an outward gesture. The sung tone depends upon the drawing of the cords together so that they can vibrate and set the air in motion. This antagonistic system can be more or less efficient. The better the cords adduct (in terms of efficiency, not sheer strength), the better the sound (in terms of efficiency rather than what is perceived to be beautiful). The efficient drawing together of the vocal cords depends not on the volume of air taken in or squeezed out but on what constitutes efficient breathing in mechanical relation with the larynx. The Italian term '*inhalare la voce*' derives from an understanding of this seemingly contradictory breathing dynamic. It may be useful to remind ourselves that we *exhale* prior to activities demanding strength of our arms and upper chest. In lifting or banging something, for example, the vocal cords close of their own volition. In this case, remember:

1. We have not *taken* breath; we have at least partially emptied our lungs
2. This automatic response tells us that our vocal cords can close strongly unaided by pressure of breath

3. In singing, it is breathing outwards from neutral (before we overfill our lungs) that facilitates the necessary conditions for this breathing out/breathing in antagonism.

The fact that breathing out increases the desire to breathe in means that at whatever point our cords close on our breath's outward journey, they do so with a degree of resistance (not resistance to an outward force of air but muscular resistance *against closing*) – they close wanting to open. This is an entirely healthy antagonism of breathing forces. This simultaneous out-in system is significant in communicative as well as physiological terms. In singing, our intention to express ourselves – to 'express life in sound' – induces precisely this out-in antagonism. We cannot therefore dissociate the singing voice from efficient breathing.

Seeing that the postural muscular frame facilitates strong, extensive exhalation, and that this in turn induces a strong reflexive response from the muscles of inhalation, we already have half the picture of the singing apparatus in view, and have paved the way for an important adjustment in thinking about how the singing voice is set in motion.

The stretching and tensing, opening and closing of the folds determine with the utmost precision how much of the folds vibrate in length and bulk at any given moment, thus giving pitch, volume and quality to tone, without any appreciable extra effort or quantity of breath. This system (a valve with astonishing complexity) is the other half of the picture of the singing apparatus. The two halves are interdependent in terms both of operation and expression.

The refined calibration of the muscle fibres to be found at the margins of the folds is crucial to the integrity of the singing voice. It is responsible for the greatest achievable vocal refinement: the *messa di voce* (mixed voice), the full-voiced *pianissimo*, the true *mezza voce* (half voice), as well as the finest tuning and agility. It is widely acknowledged that such facility is hard won. What may not be so well known is that this is because achieving the finest results (and sustaining such precision within the larynx) requires the greatest overall strength of the instrument.

We could see the singing voice as reflecting a critical meeting of opposite, mutually dependent and balancing human qualities: on the one hand physical strength and, on the other, attributes of the developed intellect such as artistic and musical sensitivity. In singing, the one without the other is found seriously wanting. For best results we must rebalance our human sophistication with our primitive nature. In training this can be linked to specific, highly differentiated muscle work (some meticulous, some relatively crude) in the process of re-integrating aspects of the voice which, because of quite contrasting characteristics, may appear on the surface to be unrelated.

So far we have considered four general areas important to singing: the aural (voice-monitoring system), the intellectual (speech), the emotional (feelings) and the physical (the body). We have viewed them almost as separate entities. They have often been accepted and treated thus because of the normal condition of the voice, which demonstrates its own lack of coordination, especially when it wants to sing. A voice does not need to be in a pathological condition to be recognisably phonasthenic, or 'unable to sing'.

We have also recognised three areas of physical vocal activity as such: the larynx, the breathing system, and the extrinsic laryngeal musculature that links the two. When we make these kinds of connections, we begin to understand the difficulties in defining the singing voice. Lay the parts of any other musical instrument out separately and they may or may not be recognisable for what they are. But the violin, for example, can be said to be a violin only once it has its permanent violin shape. The various parts of the singing voice cannot in normal circumstances be seen, let alone laid out, and cannot be stuck together so that we can then say incontrovertibly, 'that's the singing voice'. The singing voice is 'constructed' only at the moment of singing, and, unlike the violin, is only recognisable for what it is when in action.

The web of life

In *The Web of Life,* the scientist and philosopher Fritjof Capra writes,

> The *pattern of organisation* of any system, living or non-living, is the configuration of relationships among the system's components that determines the system's essential characteristics. In other words, certain relationships must be present for something to be recognised as – say – a chair, a bicycle, or a tree. That configuration of relationships that gives a system its essential characteristics is what we mean by its pattern of organisation.[7.1]

It is the configuration of relationships (how the parts relate) that counts. In this sense the voice is defined by its wholeness, not by its parts or even by its structure. It is only when a voice is whole that its full range of attributes or parts, both universal and individual, are fully complementary and effective. In revealing this pattern of organisation, our 'aural picture', the singing voice defines itself.

I have described how healthy posture makes for healthy breathing, and how breathing can help us maintain good posture. Breathing and vocalising are also interdependent for their health. The interrelatedness and mutual dependency of these various physical zones is a natural design which contributes to healthy living on the structural plain, and maximises our ability to be emotionally expressive in sound.

The efficiency of the larynx demands efficiency and strength from those muscles in the throat supporting it. With this mutuality of concern the voice is enabled not only to respond in a spontaneous manner, but accurately to our perception of musical pitch and rhythm, tonal colour and varying speed and volume, without force or contrivance. It is able to *reflect* emotion directly and appropriately without simulation or deliberation.

The singing voice's mutually dependent and defining forces operate antagonistically: examples include breathing out versus breathing in, elevation versus depression of the larynx, closing versus opening and stretching versus tensing of the vocal folds. This 'mutual opposition' provides the basis of *balance*, or, to put it another way, the simplest form of mutual 'support'. If the opposing forces are matched in efficiency and strength, compensatory struggles are no longer necessary.

If we consider that the whole vocal system is made up of antagonistic forces, we realise that 'support' as commonly applied to the voice, is something of a misnomer. Each element of the singing voice is equally important in determining both the balance of the whole, and variations of position or emphasis within its domain. The vocal organ has in fact the nature of an organ-*ism*: allowed to be itself, it is *self-supporting*. This is the meaning behind 'holistic'.

This view of the human voice takes on great significance when we examine singing in the 'classical' sense, because it suggests that a true *legato* (considered by the Italian School to be the basis of the best possible musical vocalisation) is not merely a linear matter. Inasmuch as a sung phrase should be something essentially seamless, *legato* can be perceived and appreciated *laterally*. What goes into making the perfect linear tone also facilitates continual dynamic shifts in the tone's vocal, musical and emotional content. If a *legato* sung tone was able to be cross-sectioned at various points you would see (better still, hear) at each of them a different mix of tone colour, volume, emotional weight and so on.

Facilitated movement involving the whole vocal/emotional complex turns out to be highly sensitive and responsive to a singer's imagination. Therefore, a state of expressive tonal flux is created. The intimacy between our creative expressive intention and its manifestation in spontaneous sound is perhaps what best characterises the true *legato*: like flowing water (fleeting tone) it adjusts itself to the terrain (musical contour and text), so that potential or perceived hazards are taken in with the flow. A well sung phrase has the quality of unassailable inevitability, which cannot be contrived.

True support, or 'self-sustained flow', begins to break down if one ingredient of the expression-in-sound continuum weakens, and effort becomes misplaced or misdirected. As a singer's dynamic vocal system gains in precision and staying power, so does the ability to sing a long phrase ('*il tono lungo*') in one breath, without undue effort or conscious muscular interference ('*senza muscoli*'). Balanced muscular strength prevents unnecessary expenditure of energy by convincing the body that it need no longer push nor hold back.

The vocal precision and endurance required of a voice in good singing are usually only achieved through a specialised process of re-training. Manipulative measures to effect vocal colour, volume, pitch, emotion, rhythm and verbal articulation can jeopardise the singer's chances of ever achieving an unselfconscious and stable state of vocal flux.

PART II

Chapter 8

Mixed Objectives

Perhaps the most obvious justification for the *teaching* of singing is the musical one. Western classical music is formulated in such a way that it needs to be intellectually understood before it can be 'made'. There are questions of language, style, interpretation and musicianship to be considered. To the would-be singer these are aspects of singing which do *not* come naturally and therefore must be studied and learned. So far, with the additions of language and text, this parallels the path of an instrumentalist, who usually and logically becomes a musician and 'player' simultaneously. However, there's a difference between a good musician per se and someone who makes music by singing it. Strange as it at first may seem, real problems can and frequently do arise when in the early stages of voice training it is insisted that singers learn their musicianship through singing. The reason is simple – so much depends on the state of the instrument. The singing voice is fundamentally different to any other musical instrument. While a good musician or adept performer *might* make a meaningful, presentable job singing with an inadequate instrument, unless a singer's musicality and his voice (his instrument) match up, each suffers from the shortcomings of the other.

Along with having to satisfy complex musical requirements, the singing voice is the instrument through which human beings most clearly and directly sound their feelings. At the same time this singing voice is partially used for speaking, a fundamentally different vocal activity. While a skilful, expressive instrumental musician and his instrument may seem to be 'at one', they are not bound up with each other's emotional physiology – a musical instrument has no such thing! In singing, however, the vocal and the emotional physiology are effectively indistinguishable. A musical instrument has no organic life of its own, and remains a physical entity whatever the mood or intentions of its player. It is also easy to maintain it in good working order. By contrast, the singer's voice must satisfy various roles: articulator of text, conveyer of emotions (directly), physical performer and musical instrument simultaneously. It is easy to see potential conflict between these different fields of expression. Furthermore, the singer's instrument is subject to the physical and emotional condition of the singer.

Unlike an instrumentalist's technical study, a singer's vocal study rarely begins before his late teens, by which time his multipurpose instrument will have been in part neglected and in part maltreated (for reasons already described) most of his life. A singer's instrument is shaped and reshaped because *his* life has been *its* life. In most cases a student singer's voice is far from being in good 'playable' order.

As teachers, we fail singers and singing as an art of deep human significance if we ignore the fact that the singing voice requires a radically different treatment from any other instrument. Even a basic understanding of the physical nature of the voice makes it clear that both the content and the timing of the training process must be specially considered, from the need for general fitness to the apportioning of laryngeal muscle training. *When* a pupil is ready to sing, *what* she sings, how particular she is with *how* she sings it, and *for how long* she sings at a time take on greater significance when it is understood how these factors reinforce or can impede the training process. Clarity regarding the practical differences between the process of training a voice and its employment in making music is of the utmost importance for teachers and singers alike.

Under the inadequate teaching conditions that widely prevail the singing teacher is faced with a double-bind: she or he is obliged to get the aspiring singer to make music with an instrument ill-equipped to do so. The instrument being in certain respects unresponsive, there seems no alternative but to devise methods of satisfying the demands of the music. Music invariably requires vocal agility, musical phrasing, clear text and unfettered (albeit appropriate) emotional expression. The piece in question may also venture into areas of pitch and dynamics with which the singer is uncomfortable. Thus, the singer is encouraged to 'play' on an *incomplete* instrument which keeps threatening to collapse or seize up. As a result, *forte* may be forced and strident, and *piano* weak and pallid. The sung sound not yet fully re-formed cannot flow but must be helped from one note to the next. Undue effort, stiffness or hesitation in the emission of the sound make attempts at emotional expression seem contrived or even absurd. As for words – they usually fail!

While it is correct to think that there's nothing the singing voice likes more than singing, it is naïve to suppose that a faltering voice will adequately right itself through constant use, any more easily than a dislocated leg joint will right itself through walking. Voices vary considerably in the seriousness of their faults, but these need to be addressed for what they are – vocal. Some of the greatest singers have eventually suffered from the failure to recognise and deal with early vocal faults or weaknesses. Giuseppe di Stefano is a classic case; the gold of that youthful voice (in his early twenties) showed not only its promise but also the impurities which eventually destroyed it.

'Technical' work on a voice is a process of investigation accompanied by rigorous and meticulous labour. Only after his voice has been sufficiently unblocked and strengthened can a singer be expected to sing music with ease and articulate words meaningfully, and only then can his musical merit be fairly judged through his singing. Meanwhile, by singing with a faulty instrument, a singer runs the risk of *compounding* his vocal problems: he adopts a 'singing mode' that, at least for a while, 'works'. This consists of both good and bad elements doing their level best to get on together but actually getting in each other's way. The singer is simply practising and exercising his current muscular configuration whilst endeavouring to produce musical or expressive as distinct from vocal results. As long as we sing we tend to consolidate whatever it is that's physically producing the sound, whether it is physiologically correct or not. Evident strengths, often perceived in isolation, are highlighted and promoted, while weaknesses are left to languish. A violinist will not play on a violin the parts of which do not work together or support each other. A singer should not be expected to 'play' on such an instrument either.

Creating strategies with the objective of surmounting or disguising vocal dysfunction (as distinct from eliminating it) delays a singer's rehabilitation. Such approaches are evasive, and 'improvements' are superficial and short-lived.

Singing will only improve a voice when it can be said to be in a good 'singing' condition, with no serious imbalances with which to contend. During the process of reaching this level of responsive fluency, when and how much actual singing is done must be judiciously monitored. Furthermore, singing objectives must be realistic and clearly stated. Timing and apportioning of different types of work on the voice varies greatly from person to person (as well as from session to session). Vocal facility and strength generally lag far behind what is expected of singers in terms of sensitivity to text and music. Artifice is often rife at all levels – among experienced singers who've *lost* fluency as well as students who haven't yet found it.

The principle and logic behind the avoidance of mixed objectives need to be understood. The natural fluency of the sung tone (vocal *legato*) is the *sine qua non* of effortless and truly effective singing. Therefore at any stage the process of freeing a voice might demand that the strictly musical as well as the verbal elements in music be held in abeyance for a while. How to vocalise and how to sing (with all that singing implies artistically and humanly) are very different, but generally confused. Teaching as a means to an end is commonplace and generally causes problems rather than solving them. The fact should be faced that there is a lot of singing which is far from satisfying linguistically, aesthetically, emotionally or musically.

Ironically, it is often the more enthusiastic singers who suffer most, flexing their muscles on repertoire as yet too heavy or too complicated for them, or indulging in long bouts of mindless vocal 'practising'. A singer who wants a satisfying career, making the most of his vocal assets, is obliged by the nature of his instrument to liberate it and maintain it in good condition, husband his vocal resources and be judicious, even restrained, in his choice of repertoire.

The process of *un*-mixing objectives

We might not *know* how to sing Mozart or Berg, for example, but on some ancient level of our being we know *how to sing.* If this were not the case, we would not have been provided with a singing voice and a strong desire to use it. If we cannot sing it's because our voice has fallen into disuse or disrepair. Singing is a spontaneous and creative act only insofar as the voice is *able to respond* to the impulse to sing and our imaginative intentions. The job of voice 'trainers' (as distinct from musical coaches) is therefore no more nor less than to guide the re-establishment of the natural singing condition.

If the breakdown of a voice has been the result of a lifelong inhibiting or conditioning process, regaining its functional integrity will require a freeing or reconditioning *process.* Unfortunately, this concept, logical as it may be, is unlikely to figure with a person whose voice is somewhat unresponsive because the physical information he's receiving suggests that he is '*doing* it incorrectly'. Illogically he will use will-power sooner than admit inability, force sooner than spend time exploring or patiently honing his instrument. His concern and determination will only exacerbate his difficulty and reinforce his false assumption. Taking the analogy of manual tools, a blunt or rusty saw will make us sweat and hack, whereas, if it is sharp and in good condition, employing it effectively requires only direction and an easy, rhythmical swing of the arm. Many singers struggle with their voices simply because they are 'blunt and unwieldy tools'. This causes confusion between what the mind thinks it can or ought to *do* and what the body *asks for* by way of rehabilitation. This vicious circle is perpetuated by teaching that concentrates on what is '*wrong*'. Precious time is often wasted in attempts to break the circle with un-physiological short cuts to imagined goals.

Frustration sets in when our intention is 'right' in our head (the appropriate emotions are felt, and the imagination is vivid) but our voice doesn't respond as anticipated. 'I keep *getting* it wrong' may simply be because 'I keep trying to *make* it right'. But can we really behave as though there are such things as 'accurate emotions' or 'correct fluidity'? What options are there apart from effort and artifice?

To ensure that a genuinely progressive process is evolved we must break away from the 'product' mindset and instead consider:

1. It is unrealistic to expect specific evidence of singing attributes from a voice which is blocked, unbalanced or lacking adequate strength.
2. What a singer *has* is more important to the process than what he lacks.

3. We must find systematic and adaptable ways of liberating voices, their individual qualities, and their inherent lyrical and emotional natures, and of developing individual potential, encouraging these qualities and not just the 'mechanics' behind them.
4. The training process must aim in the most thorough way to prevent future backsliding.
5. It is the freely working or physiologically well-founded voice that truly 'sings': has what we call a 'singing quality', has the musical skills required in vocal music, and can convey what is intended in emotional and intellectual terms because these are the capacities with which the singing voice *is naturally equipped.* If we don't acknowledge and trust the *design* of this voice, our efforts will always be misplaced.

Even given the best conditions and intentions, frustrations and feelings of inadequacy from lack of achievement can set in early in a singer's development. Profound discomfort can arise from the seemingly irreconcilable objectives of performance on the one hand and 'becoming a singer' on the other. A knowledgeable and sensitive musician can go through hell if his voice is not yet ready to 'play' with the spontaneous ease that he so clearly imagines. However well they may eventually complement each other, being musically literate and being able to sing are very different.

The various aspects of singing must not be jumbled together at the outset in the vague hope that they will sort themselves out. The trainee needs to be enabled to concentrate on different aspects of the work – physical, musical, textual or emotional – while gradually and consistently being made aware of how the various strands will eventually draw together and interweave. In Chapter 2 we saw that not only does normal speech often conflict with the singing voice but the singing voice is often at odds with itself. This, paradoxically, is because of what singers feel is expected when they sing. Insistence on clarity of diction while the emission of the vocal tone is still impeded, for example, may cause a conflict of interests. Resolve the conflict and we see that not only can words and voice complement each other in the cause of dramatic and emotional expression, but that they need each other if communication is to be unambiguous.

As long as we're thinking 'how to do it' while we're singing we can be neither good vocal musicians nor clear communicators. Different working agendas are indicated in restoring the vocal instrument to its natural working order, and making music with it. While a voice must be tested during the course of its development against what we are expecting of it in expressive terms, we must guard against premature insistence on 'getting right' those aspects of the eventual performance which are mind-directed. What *can* be and is being achieved must be objectively assessed at all stages and shown to be helpful to solid progress. The voice itself will tell us both how a singer is progressing and what can reasonably be expected of her singing. In addition, we must understand the interdependent nature (as distinct from expedient interrelating) of language, sung tone, emotion and musicality in singing.

The shortcomings of a singer's singing are usually revealed to be primarily of a physical or physical-emotional, rather than a musical or linguistic nature. As explained, our primitive self has suffered a process of subjugation. We have lost touch with or overruled our instinctive nature and thereby lost the benefits of spontaneity. Singing is only an intellectual pursuit inasmuch as it is to do with music and text. The vocalising which serves these things is instinctive.

In training, having a go – without thinking – is almost always more fruitful than thinking it out first. As a child I was probably cautioned about the rashness of running across the seaside rocks (the dry ones!), but as an adult I have observed the ease with which children (and dogs) manage this feat. A voice that is not sufficiently liberated and is preoccupied with what it has to do and how to do it will always be hesitant; it will always be fearful of the rocks and make inhibiting, cautious efforts to negotiate them. The voice is only truly up and running when we can trust it to follow our imaginative intentions. Singing is then no longer a technical matter but a spontaneous response to music and the joy of making it.

PART II

CHAPTER 9

Vocal Liberation

Towards a definition of vocal liberation

What is 'liberating' or 'unlocking' a voice, as distinct from the act of singing itself? Does the speaking voice, trained well enough, become the singing voice, and is a trainee actor, therefore, making part of the journey towards singing, and the trainee singer travelling the whole way? At what point in the process is either's training complete? I have illustrated the differences between singing and speaking, and it is clear that, when a person attempts to sing (either formally or quite spontaneously), what he's doing isn't speaking. It may not be fully singing either, although recognising potential we might suggest 'You should get that voice trained.' It's *that voice* with which the person is attempting to sing, not the one with which he habitually speaks.

I've explained that the voice that attempts to respond to our *desire* to sing is *different in structure* to that of normal speech. This singing instrument might be trained to make *any* sound, and indeed there are many sounds that we call collectively 'singing', just as there are many sounds that are called speech. Do we as teachers 'mould' a voice into, for example, a jazz or operatic singing voice? If a voice seems to possess attributes making it *suitable* for one of these genres, do we try to make it more jazz-like or more operatic, according to sounds we associate with these styles? It is easy for a teacher to be seduced by the idea of nourishing the next Billie Holiday or Edita Gruberova! Should we train a singer differently for Early Music, lieder or opera? Should a voice be trained to perform specific skills, feats or effects associated with different kinds of music?

The negative consequences of specialised aims of this kind are widely evident. Can we justify limiting a person's capacity to express his or her true self? Surely the role of a responsible voice teacher is to train a voice in accord with its nature so that it works well in all respects.

Healthy design

One current school of thought would argue that there are separate 'well-workings' for opera, jazz, musical theatre and other recognised styles. It's obvious that some ways of vocalising, in any style, are more effective and less harmful than others. However, there must surely be degrees of vocal health per se, just as there are degrees of general physical good health. The question 'what constitutes a healthy voice?' is the ongoing concern of many voice specialists just as what constitutes a healthy mind is the concern of psychologists. Teachers of singing should be more concerned about vocal health than about what sounds 'operatic' or 'lieder-like'.

It's misleading, if not dangerous, to assume that a singer's voice is healthy because it survives or copes with a style of singing with little discomfort, even if his vocalisation is effective and deemed to be 'safe'. There is all the difference between being effective according to some given criteria, whether 'scientific' or musical, and being in good health. In life, 'surviving' and 'living' constitute entirely different attitudes and aims. Training can lead a voice away from its original expressive nature while it remains effective musically. On the other hand it can lead the voice to fulfil its original potential, so that it truly lives, a genuine expression of its owner, and an indicator of his or her well-being.

Putting styles of musical vocalisation aside, we can measure vocal health by natural rather than manmade criteria. Training can be a process of improvement in vocal health regardless of what we want to use our voice for. This may not appeal to singers who wish to imitate a particular style or 'model', but it makes sense to those who see their voice as primarily an expression of them*selves*. Too often a trainee singer says something like 'I want to be a soprano like Callas!' It's important to remember that while singers of various styles have become well-known for their own vocal qualities or peculiarities, these have always been unique. It would be perverse to suggest that it's a good idea (let alone a healthy one) to cultivate strange, limiting or degenerate vocal habits on the off-chance that they'll make us famous. Emulators of Elisabeth Schwarzkopf, Maria Callas or Peter Pears (to cite three examples of well-known singers with quite distinct vocal personalities) have rarely succeeded in achieving anything more than poor imitations. That their models became famous is a measure of their personalities or considerable artistic merits.

If singers specialise early on (in Baroque music or musical theatre, for example), they run the risk not only of limiting what they can 'say' with their voice, but of straying from their true vocal and artistic personalities. In medical parlance 'healthy' means sound, and 'holistic' means treating the whole person. Healthful and whole share the same root. A complete voice is bound to be the one that serves us best for whatever purposes it was designed.

Training

Like many of the words used in conjunction with singing, 'training' has unhelpful and misleading connotations: animals are 'trained' (to be domesticated or perform various feats), and athletes train to be super fit. Neither example is helpful applied to voice training. Inasmuch as a singer's voice is in disarray, it must be regenerated or reorganised in its entirety before it can become fit for the activity of singing professionally. In this respect singing is probably quite unique. Unless the physical and physiological connections between a singer and her voice are understood, objectified and addressed for what they are, training in the generally accepted sense of the word can do a voice more harm than good. Attempting vocal 'feats of strength', distorting or unbalancing the whole, or capitalising on strengths at the expense of continuing weaknesses will only create more divisions or emphasise existing ones between the singer and her true sound. As well as the obvious negative consequences for communication this can have serious psychological (as well as physical) implications.

An individual journey

Regaining our voice is to some extent a reversal of what impeded its progress in the first place. Training is, therefore, a holistic reunification of the singer with himself. This is particularly important in the classical arena, where the greatest vocal and emotional diversity are required of the performer. Training is not a question of cultivating individual vocal peculiarities or propensities, but *is* an individual matter. As well as sensitively dealing with a singer's vocal (as distinct from musical) shortcomings and habits (probably acquired over a long period) teachers are responsible for nourishing each individual's special vocal or tonal quality which may not be easily discernible at the outset.

In attending to a singer's natural vocal coordination we are honouring her potential as a unique communicative being. The process is continually reinforced by the building and employment of natural strength and stamina which gradually develops into the ability to sustain the voice intact, maintaining its flowing freedom, for the duration of a song recital, operatic production, or long hours of rehearsing, each of which puts its own specific demands on the performer's fitness. Fitness and stamina for singing, as distinct from boundless energy, can only be acquired through thorough training.

Different in kind

Singing and 'training to sing' can seriously hinder each other's progress. On the one hand, we can be struggling in performance to employ our musicality and understanding of text and drama with an unwieldy instrument; on the other, it's easy while training to be side-tracked by irrelevant intellectual issues.

Logically, the physical instrument as such must command the greater part of a singer's energy until it begins to serve his singing instead of 'getting in its way'. Assiduously adhered to, a flexible but rigorous and systematic training process will eventually lead to clear articulation, authenticity, equality of tone and strength throughout the singer's natural vocal range. Such training is rarely straightforward because of its personal nature. Our voice embodies and reflects our whole being: body, mind, heart and soul.

Dedication

If training is to be effective it requires the whole-minded, whole-hearted and whole-bodied commitment of the singer. Many talented would-be singers are prepared to commit only to a certain point. Invariably they make little or painfully slow progress relative to their potential. The physical-emotional parts in singing are commonly the most difficult to access or submit. Bodies are generally lazy and emotions timid, even in singers relatively free of tension. Singers often try to work from a 'mental control tower', at a comfortable distance from where the action should really be. Significantly, the most reticent department in our work, our emotions, is ultimately the one which makes singing such a potent form of expression. A teacher must learn to recognise individual 'holding-back' strategies and exercise appropriate patience and humour while reticence is gradually overcome. No one element must be allowed to 'run away with the show'. Only by degrees can the various components come into their own and cohere in their naturally balanced proportions.

Radical treatment

The more knotty or deeply ingrained the problems of a voice, the deeper and stronger the treatment called for in its regeneration. The teacher is often faced with a voice at odds with itself – the wrong kind of antagonism! How we begin to work with this state may be crucial. Trying to treat the voice as a whole when its parts are out of tune with each other can only result in a technique in which some elements of the voice will remain stronger than others. Concentration on the most obvious attributes of an individual voice (fluency in coloratura for example) while failing to seek out those which are more obscure also results in an unhealthy one-sidedness and continuing instability, both physical and psychological.

To train a voice thoroughly a teacher has no choice but to hold in view and in balance three ideas:

1. The singing voice is a natural entity
2. Radical treatment is often indicated because of the deep-rootedness of certain problems, or the incompatibility in strength between different vocal spheres.
3. There is clarity in separation.

Sounds strange – audible tension

Any muscular activity partially isolated from the whole will sound precisely what it is, partially isolated and *not* the whole! Deliberate exaggeration of this activity may seem to take the sound a step further still away from the whole. Such sounds might be like exclamations, or resemble the vocal expressions of animals or birds; the kind of sounds that provoke dismissive or derisory descriptions of some serious vocal training! All sounds issuing from the human throat, however imitative, are human by definition; they reflect describable muscular activity in the body, throat and larynx. A trained ear will recognise through activity-reflected-in-sound what is relevant or useful to the training process and what might pose a threat to vocal health or an obstacle to its freedom. Exaggerated contraction (a 'strange sound') can be as helpful in reducing unhealthy muscular tension as it can in increasing healthy tension.

Experiment 9.1: Relieving unwanted tension through deliberate exaggeration

If you have tense shoulders, draw them up towards your ears. Hold them there for a moment and then let them freely drop. The drawn-up shoulders feel abnormal and look quite out of place. However, the treatment is not considered foolish; it's an effective means of releasing tension.

Invisible tension

In training a voice we should not only recognise when unhealthy muscular tension is serving a 'useful purpose', but also that much of it has little if any visible manifestation. A singer's stance or the way she moves may give us a clue, but these signs don't necessarily indicate the root of the problem. Although a teacher may be able to feel a singer's tension with her hands, to make a thorough enough investigation and be able to conduct the training process holistically and accurately, she must rely on a more acute device – her ears.

Specific tension or its cause may not be immediately discernible aurally. Nevertheless, *what produces* a vocal sound will always manifest itself in that sound. An obvious proof of this is the following experiment, turning the usual process around.

Experiment 9.2

Make a tense-sounding sung sound, followed by a sung sound without that tension. The difference is immediately clear.

In training, a certain amount of judicious experimentation is required, with the understanding that we're not looking for a 'more attractive' sound, but one that makes unwanted tension unnecessary, or encourages feeble muscles to rouse themselves.

Vocal athletics

Since the act of singing is first and foremost a physical activity, the training of a voice can be compared to that of an athlete. However, there are differences, which must be understood if training is to be sufficient but not overdone. This was brought home to me when a laryngologist asked me the difference in objective between the singer's and the athlete's training. I told him that an athlete tries to knock seconds off his time and a singer tries to sing his best. The laryngologist's interpretation was even more apposite: 'In other words the singer tries to optimise and the athlete tries to maximise his performance.'

For trainers of voices, the understanding of this distinction is of fundamental importance. I've met singers who *are* out to break records (or glass, come to that!) but in my experience conscientious young singers want to sing as well as they can, not necessarily louder, higher or longer than anyone else, and relish the applause for their singing rather than their physical prowess. For athletes (whose aim is to be better than others) competition is essential but for singers it can be damning, leading to misplaced effort and the distortion of values.

Being in poor physical shape is bound to cause poor singing – relative, that is, to the individual's capability. A singer's performance will always depend to a certain degree on his fitness and to this end he trains physically. To be an effective trainer we must examine what this aspect of a singer's development means in the broadest and most detailed terms.

What it takes

Athletes submit to rigorous training programmes, and expect to make appropriate sacrifices in other areas of their lives. While the whole ethos may appear different on the surface, there is in fact the same necessity for a professional singer to be in good physical condition, if only because, like the athlete, she *is* her instrument. Even without competitive pressure, singing is an amazingly demanding job, requiring mental and physical discipline and stamina. A singer should aim to achieve peak fitness for her activity. Many unknowing people assume that singing is a 'soft option' career – you just open your mouth and sing! Aspiring young singers cannot be blamed for having little conception of what becoming a good vocalist and interpreter entails. It's the responsibility of teachers and institutions to provide prospective professionals with a realistic picture of what training for their chosen career might mean.

For some, the concept of hard physical work applied to singing is an unpalatable contradiction in terms. Remember though that we are talking about training. A good ballerina looks as though she has come into the world on her *pointes*, thus belying her many years of hard, dedicated training. Performers whose movements are unduly effortful often appear ridiculous and their awkwardness can make an audience feel uncomfortable. Ironically, this is the result of *too little* work when it *was* needed.

I have found again and again that singers, inexperienced and experienced alike, have little or no idea *how* to apply themselves in a training session when they start work, let alone what preparations to make for it. How to work effectively without mixing objectives or wasting energy must be constantly studied by both teacher and pupil.

Understanding 'the physical nature of the vocal organ' (the sub-title of the book *Singing* by Frederick Husler and Yvonne Rodd-Marling)[9.1] makes for the evolution of training which is not only purposeful but appropriate and rewarding, paving the way for music-making and voicing articulate, meaningful text. To this end it's advisable to remind ourselves constantly that the *voice itself* is neither music nor words. Given its freedom, the singing voice is what *facilitates* these things. This perspective points clearly to the relative importance of the various components of singing throughout all training and rehabilitation procedures.

PART II

Chapter 10

Training Ground

In previous chapters I've described what I believe to be behind most of the difficulties of learning to sing, so that potential students and teachers can have an idea of the kind of journey they are embarking upon in the studio. While it may be tempting to avoid personal issues, we cannot treat singing in the abstract or the voice in a vacuum. Inasmuch as we were born with a larynx, a breathing system and, importantly, emotions, we were born with an instrument with the potential to sing. If that potential has been impeded or not fully realised, we have no choice but to look at what has gone wrong, or has not gone right, and deal with it accordingly. This is why I have described the process of learning to sing as regenerative: we bring back into existence what was there at least in potential.

Prospective professional singers are normally chosen from those whose voices in their late teens show sufficient promise to suggest that they might be trained to 'sing well'. If as teachers we predict that a voice will sing a particular type of music, we have fallen into the trap of pre-judging that voice's prospects, instead of treating it for what it is and taking time to discover what it will be. One can rarely predict at the outset how a voice will turn out. Usually its true colours and dimensions are only partially discernible. If training is to have a realistic aim and be carried out in a purposeful fashion, it must be a process of *discovering* an individual voice's true quality, calibre and capability.

For many years we have been preoccupied with so-called 'authenticity'. Nothing could be more authentic than the human singing voice. If humans are born with a 'singing instrument' it must have existed long before singing – as we know it – came into being. Manmade musical instruments have been modified or 'modernised' according to the imagination of composers and music makers, but the human's voice, like the rest of his body, has remained as it always was. The individuality of each voice is preordained by Nature. A voice in training is simply one that is being rediscovered for what it is, and all voice training programs should acknowledge this. If we begin with a 'serving-music' agenda, or, worse, a 'packaging-for-the-market' agenda, we are unlikely ever to discover the true extent of an individual's vocal wealth or communicating capacity. Each and every voice is the *mouthpiece* of an individual – his or her unique identity and natural provision for communicating that identity in sound – and thus it could hardly matter more whether we set out to impose an idea on a voice or to honour its design.

I've already given some idea of the singing voice's marvellous design, but in order to facilitate a purposeful exploration we need guidelines and points of reference. The voice's blueprint is its anatomy and physiology. At this point I might provide detailed pictures of muscles and cartilages, with scientific explanations of their various roles. The operation of the singing voice, however, is something that we can only really recognise and know by experiencing it. Above all this means hearing sound. In reality we can no more hear (or even see) a voice and its current state and personal characteristics from pictures and diagrams than we can experience and appreciate the fauna and flora of even a familiar place from a map. Singing neither looks nor feels as it sounds, and singing teachers as well as singers are easily led into unhelpful, even counter-productive attempts at physical manipulation when too many muscles are on display. Anatomy and physiology books abound, as do technical and scientific books about singing and voice in general. But remember, singing is first and foremost an aural art, and teaching singing is therefore primarily an 'aural science'. Unfortunately, given human beings' propensity to imitate, even examples in sound can be unhelpful. With the minimum of uncomplicated visual aids, I hope to appeal to your imagination, since it is in this fascinating, perceptive world that we find one of our greatest travelling companions.

Zones to explore and refer to in training

When working with a voice we must determine as we go along to what extent it's in a singing condition, and to what extent and in what manner it's impaired. To these ends we can usefully recognise four physical zones between which there is normally a degree of disassociation, confusion, or outright antagonism.

1. The larynx

This being the source of the sound, we could call it the most important zone. Indeed its sounds give us all the aural information we need. However, that information-in-sound isn't just about the larynx but reflects all the other spheres upon which it depends for its efficient functioning. By listening to what a larynx says about itself, we can begin to determine what is or is not contributing to its mechanical health.

Strictly speaking, when we talk about the larynx we are concerned with what goes on inside it and what contributes to its healthy functioning in the throat around it (thus we talk about *in*trinsic and *ex*trinsic laryngeal muscles). In singing, these zones are mutually dependent. However, speaking tells us that sound can be made by using some of the intrinsic musculature with almost no extrinsic support relative to what there could be. So, useful comparisons may be made between singing and speaking because they *sound* so different. What is operational in good singing can be as clearly heard as what is absent in normal speaking.

A major difference between singing and speaking is that in the act of singing the vocal folds are lengthened and thinned. The relatively lax or otherwise tight sounds in speaking can both be attributed in large measure to the absence or inefficiency of this stretching process.

The stretching of the vocal cords is brought about by two very different actions with different, mutually beneficial purposes resulting in quite distinct sounds. It's important to know that the stretching (lengthening) of the folds and their tensing can happen independently of one another; that the folds contract as muscles do, but that being stretched they are relatively passive. The action of the ring-shield

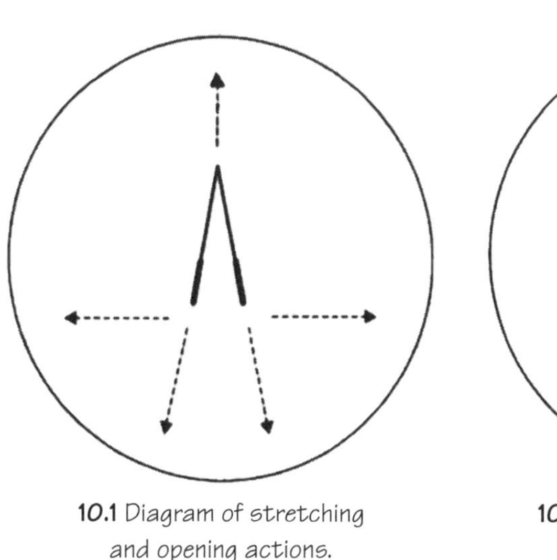

10.1 Diagram of stretching and opening actions.

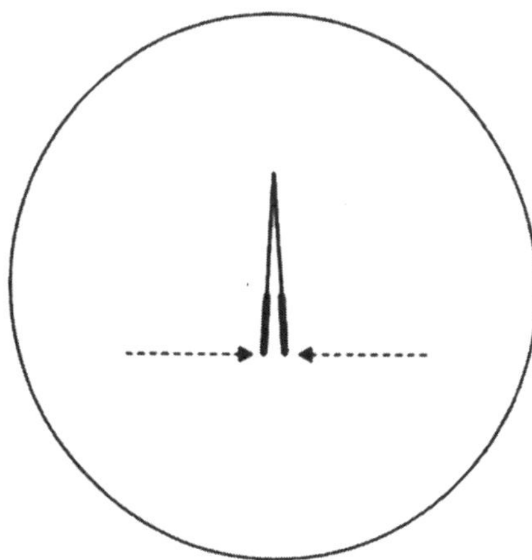

10.2 Diagram of closing action.

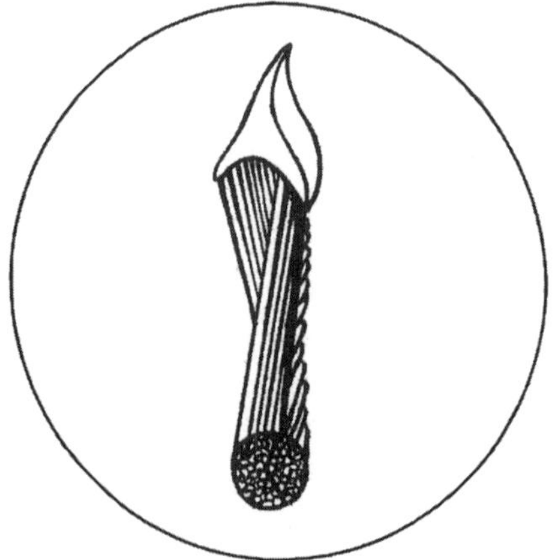

10.3 Diagram showing crossed vocal fold muscles attached to pyramid cartilage. Connections between muscle fibres strengthen as muscle bundles contract.

(After Husler and Rodd-Marling)

muscle, M. *crico-thyroideus,* although of primary importance in good singing, in fact stretches only the mucous membrane of the folds (the disembodied sound of *falsetto*). Being the lining of the windpipe, this membrane terminates at the cusp or top edge of the vocal folds which therefore have no choice but to be stretched along with it. This thinning, refining process provides insufficient resistance for much vocal fold tension. A pull (backwards) on the mass of the vocal folds themselves is therefore provided by the action of the paired *posticus* muscle situated at the back of the larynx. This strong process simultaneously *opens* the glottis (triangular space between the cords) making for the efficient intake of breath (its primary role, as in yawning). However, we are immediately aware that gaping cords do not make for efficient singing. The stretching/opening process is therefore matched (antagonistically) by an appropriate means of *closing* the glottis so that the folds can be stretched, opened (abducted) and closed (adducted) simultaneously. In this condition vocal fold tension can be introduced with optimum economy and effect to produce the singing voice.

However well the two vocal folds are tensed, opened, stretched and closed by these means, there remains a weakness at about halfway along their length, resulting in some air escaping 'unused'. Correctly developed, the membranous stretching results in a thinned edge to the vocal folds, and it is the action of the 'edge-mechanism' (the muscle fibres radiating to the folds' outermost margins) in conjunction with this stretched membrane that finally reduces the use of air to a seemingly negligible amount, and turns the singing voice into an instrument of the utmost clarity and precision. For a diagrammatic representation of the vocal fold dynamics in the course of singing, see Figure 10.4.

In other words, the stretching and thinning of the folds provides the conditions necessary for the most varied as well as intricate tensing activity of the muscle-bundles of which they are composed. We underestimate the significance of this stretching-tensing relationship at the peril of losing skilful as well as beautiful singing. While a singer may be a skilful musician and interpreter, technically it is the *finely tuned voice itself* which is full of skill. For all our ingenuity, we have to 'make it work' only so long as it is *not* finely tuned.

The complete voice with its natural pitch range and full compliment of interrelated skills entails the simultaneous coordination of stretching, tensing, opening and closing processes. In the course of a sung phrase this dynamic system is in continuous subtle change, according to split-second expressive needs. Faced with the complexity of this design, it's absurd to imagine we can get in there and 'pull the strings' or that it's possible to set everything in motion by what is usually meant by 'supporting', or by pressing the voice into action by force of breath or will. Remember how short the vocal cords are? We can and must learn to *hear* exactly what is or isn't going on inside the larynx. So long as the *extrinsic* muscles

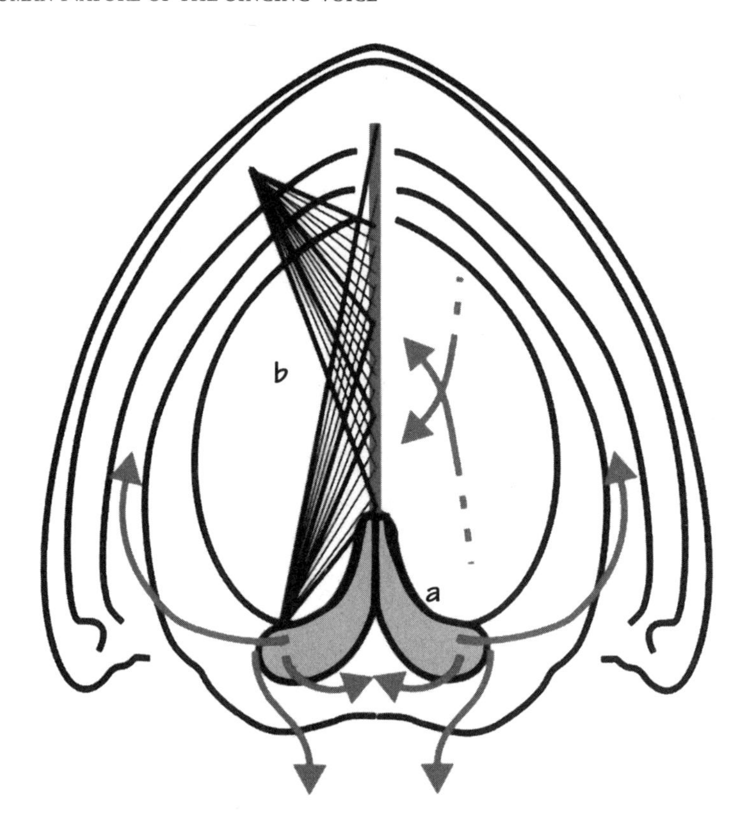

10.4 Diagrammatic representation of vocal fold dynamics in the process of singing. Directional movements of the arytenoid (pyramid) cartilages (**a**) responsible for stretching, opening and closing of the vocal folds, together with (**b**) their crossed muscle bundles.
(After Husler and Rodd-Marling)

are not fully doing their job, however, it will always feel as though some deliberate manipulative measures are called for. Such measures invariably cause more resistance than co-operation from the throat.

2. The suspensory mechanism

To alleviate the resistant or restricted throat (which may or may not feel uncomfortable) the larynx is provided with a system that 'takes the strain', giving it freedom to do its own thing without constraint or interference. The term 'elastic scaffolding' (Husler and Rodd-Marling)[10.1] may be more apt and more helpful to the imagination than the current term, 'strap muscles', because it indicates that the muscles concerned not only act to give the larynx stability but are themselves in movement, working in relation to one another, giving and taking in response to the changing and varying requirements of what they're supporting (see Figure 10.5).

Pertinent observations:

1. The various directional pulls on the larynx are flexible (elastic), as is the antagonistic nature of their relationship to each other. When they balance out in strength and flexibility, they provide the larynx with the support it needs to be able to work with a high degree of independence. In addition, the stretching and opening of the cords is assisted indirectly by these in-spanning muscles. There's no feeling of strain or tension in the throat, even when the folds are fully adducted (thus the sensation 'open throat').

2. Each of the main four parts of this mechanism is 'anchored' in such a way that in the general scheme of things this mechanism can be considered just as much part of and under the influence of posture as any other muscle system in the body, especially when (in singing) the larynx is called upon to carry on working in a body continually on the move.

3. With all the suspensory muscles being of equal importance, the larynx in singing lies neither too high nor especially low. Rather than concerning ourselves with either position we can concentrate on re-activating, balancing and strengthening this natural in-spanning process.

4. The 'depression' of the larynx (see Figure 10.5d) – forwards and downwards – acts by *reflex* to its elevation (Figure 10.5b and c) upwards and backwards (pure head voice sound). If the alignment of the larynx is to be natural and not stiff its elevation must be correctly achieved. This natural reflexive relationship is evident in yawning, an obvious throat-opening activity.

5. The additional pull backwards and downwards (Figure 10.5e) (voluminous full head voice sound) is very often thwarted by poor posture, when the spine is contracted or not fully extended. I have found this muscle to be weak in almost all beginners, as well as most professionals, even among those whose posture is generally good. It is another muscle that's rarely used in speaking. As a result, when attempts to sing are made, the larynx gets fixed by muscles above the tongue bone by way of compensation.

6. The tongue bone (see Figure 10.5f) and its attaching muscles (upper and lower) comprise an intriguing arrangement. The fact that the tongue bone is a 'floating' division between the musculature of verbal articulation and that of the sung tone may be a crucial element in singers' ability to successfully sing and speak simultaneously, using the tongue freely without its action unduly impacting on the freeplay of the larynx as a whole. This arrangement is assisted by indirect suspensory muscles linking the tongue bone with the shoulders and the sternum. These have long been recognised as

especially important for the 'formation of high notes' (G. E. Arnold quoting Ammersbach).[10.2] Drawing the tongue bone downwards by this means helps to obviate any gripping effect the tongue might exert on the larynx if asked to 'assist' with voice production when it really wants to be free to articulate words. This seems to be of increasing importance the higher in our range we venture (see Figure 10.6). The auxiliary suspensory muscles ensure that the head needs no special position or adjustment for high notes, since they reinforce the larynx's stability.

7. The soft palate (or velum) is the cause of considerable misunderstanding. In singing we don't raise the soft palate: it is raised by the action of the elevators of the larynx. We must understand this difference in order to avoid misplaced attempts at raising the soft palate which result in stiffness, as in a yawn taken beyond its onset or 'politely' stifled (when the tongue also stiffens) or as in swallowing when the soft palate contracts. (Check these out for yourself.)

8. The soft palate can play a defining role in singing when left free to do so. The in-spanning of the larynx (described above) lends to a voice a strong in-breathing quality. In other words, it is closely related to the diaphragm. I have observed in the most expressive singing that the soft palate acts in sympathy with the diaphragm, 'mirroring' its most subtle movements and thereby translating emotion into sound. This is one way in which the need to *demonstrate* emotion is obviated. Turned around, this means that if the larynx is falsely or stiffly elevated (by tongue or swallowing muscles) deliberate attempts at expressing emotion sound forced or contrived, and the possibilities of subtlety are severely reduced.

3. Breathing

When we relate breathing to singing, it's important to make a distinction between what constitutes the breathing apparatus and the breath itself. It is a popular misconception that we sing by driving the voice into action by force of breath – a kind of bellows or wind instrument effect – and therefore studios across the world talk as much about the importance of breath as about 'breath management'. The (reasonable) assumption is that if breath plays such an important role in singing, the more we have the better. It's equally reasonable to deduce from this that the greater the quantity of breath, the more skilful we need to be in managing it. Both these conclusions can easily lead to constitutional weakness in the breathing apparatus as well as struggle and stiffness in the vocal system at large.

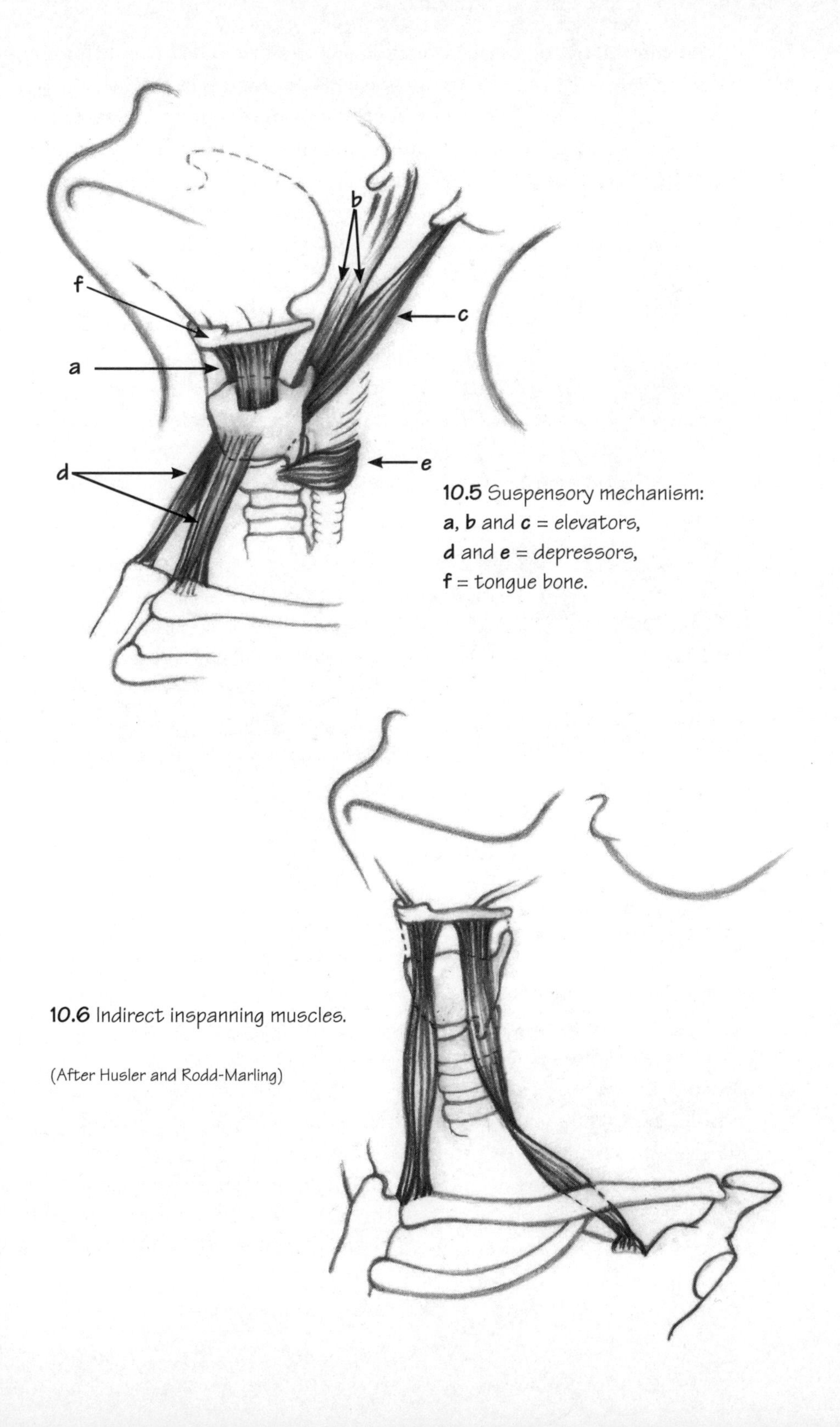

10.5 Suspensory mechanism:
a, **b** and **c** = elevators,
d and **e** = depressors,
f = tongue bone.

10.6 Indirect inspanning muscles.

(After Husler and Rodd-Marling)

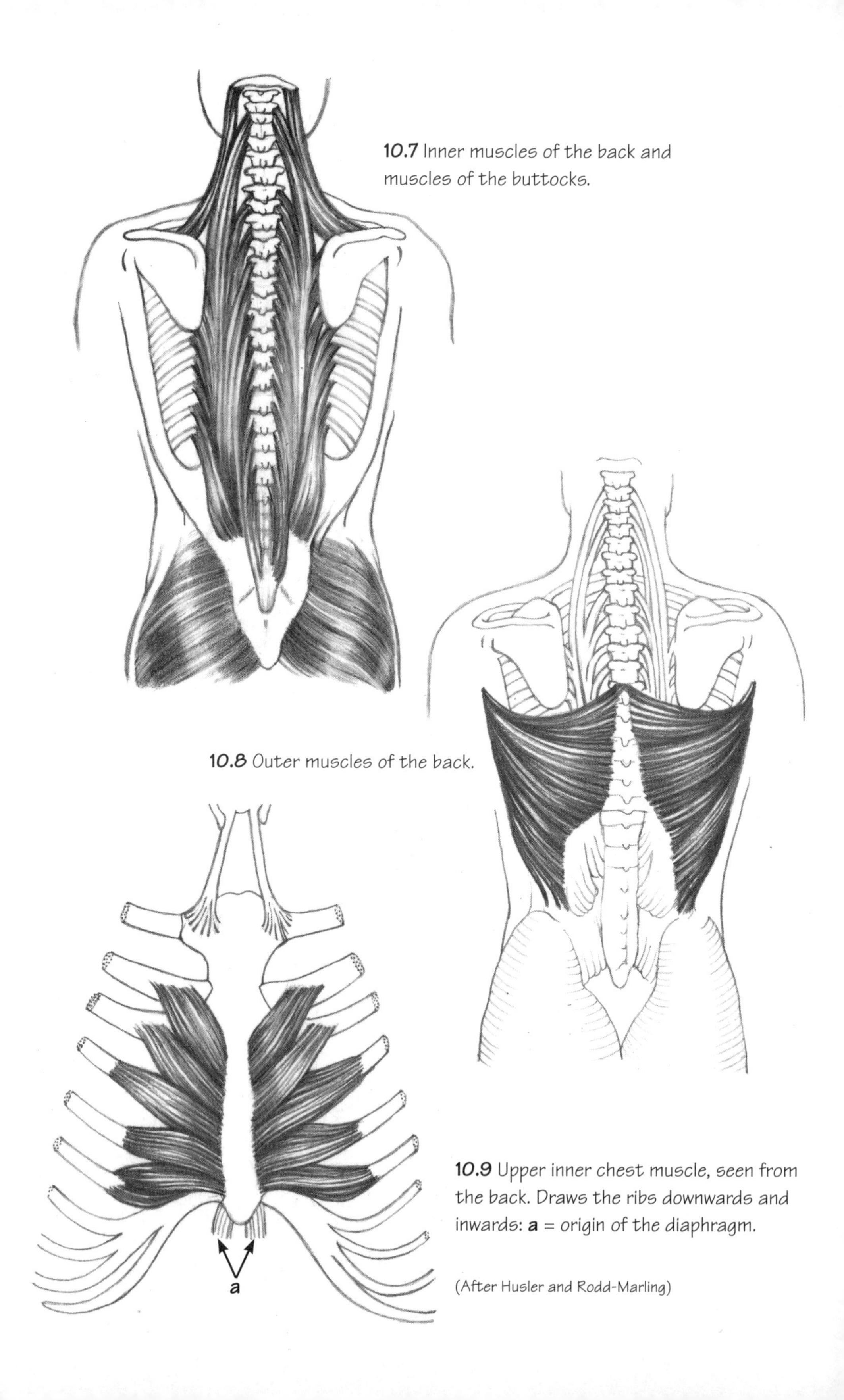

10.7 Inner muscles of the back and muscles of the buttocks.

10.8 Outer muscles of the back.

10.9 Upper inner chest muscle, seen from the back. Draws the ribs downwards and inwards: **a** = origin of the diaphragm.

(After Husler and Rodd-Marling)

Let us refresh our memories with regard to breathing in general:

1. Even if our only purpose in life was to get rid of carbon dioxide from our lungs and replenish them with oxygen, we would find that the most efficient and effective way of doing so was to breathe out to the limit of our capacity. In this way we would be able to *empty* our lungs of CO_2, so that the resulting influx of oxygen would reach and replenish the outermost bronchioles of the bronchial tree, and thence be conveyed into the blood stream. This thorough system of gas exchange is born out by the number and size of the muscles of expiration and even the complexity of the lungs themselves. (see Figure 10.10)

2. When we need to employ strength in our chest and arms we don't take in breath; the cords close to prevent this happening. This strong body-throat connection is highly significant in singing, while, by contrast, the equally strong body-throat connection caused by pressure of breath prevents the free play between these two muscular zones.

3. In training a voice we need to highly stimulate and motivate our breathing apparatus, not for the amount of breath we can gain or manage but for the apparatus's muscular flexibility and cooperation with the throat. In particular we need to pay attention to the reflex, or imperative, of breathing in by which breath is regained as the natural consequence of exhalation.

In all three cases above we see that it is breathing *out* that is important. Unsurprisingly therefore, we have an extensive breathing system for this vital activity. Assigning one part of the breathing system, such as the abdomen or diaphragm, to singing is misguided and bound to cause antagonism of the wrong kind as well as compensation, because by a very long way it is incomplete.

Powerful muscles of exhalation are to be found in the back, as well as around the sides of the ribcage and behind the sternum (see Figures 10.7, 10.8 and 10.9).

Together with the over-favoured abdominal muscles this extensive mechanism effectively means that the whole ribcage is on the move when we breathe out to any extent. Deeper layers of muscle contribute indirectly to this process. They come into their own if the exhaling movement is taken far enough.

Much as it may be necessary to regain its efficiency and strength, breathing is a natural movement, and the less we try to interfere with parts of it the better. It may be worth pointing out that going deeper, taking breathing further than normal, is usually a gradual development. As with all muscular exercise, extent as well as flexibility is gained through a process of contraction and relaxation. It's never a matter of mindless puffing and panting, bellows-like movements or breathing out and in by numbers!

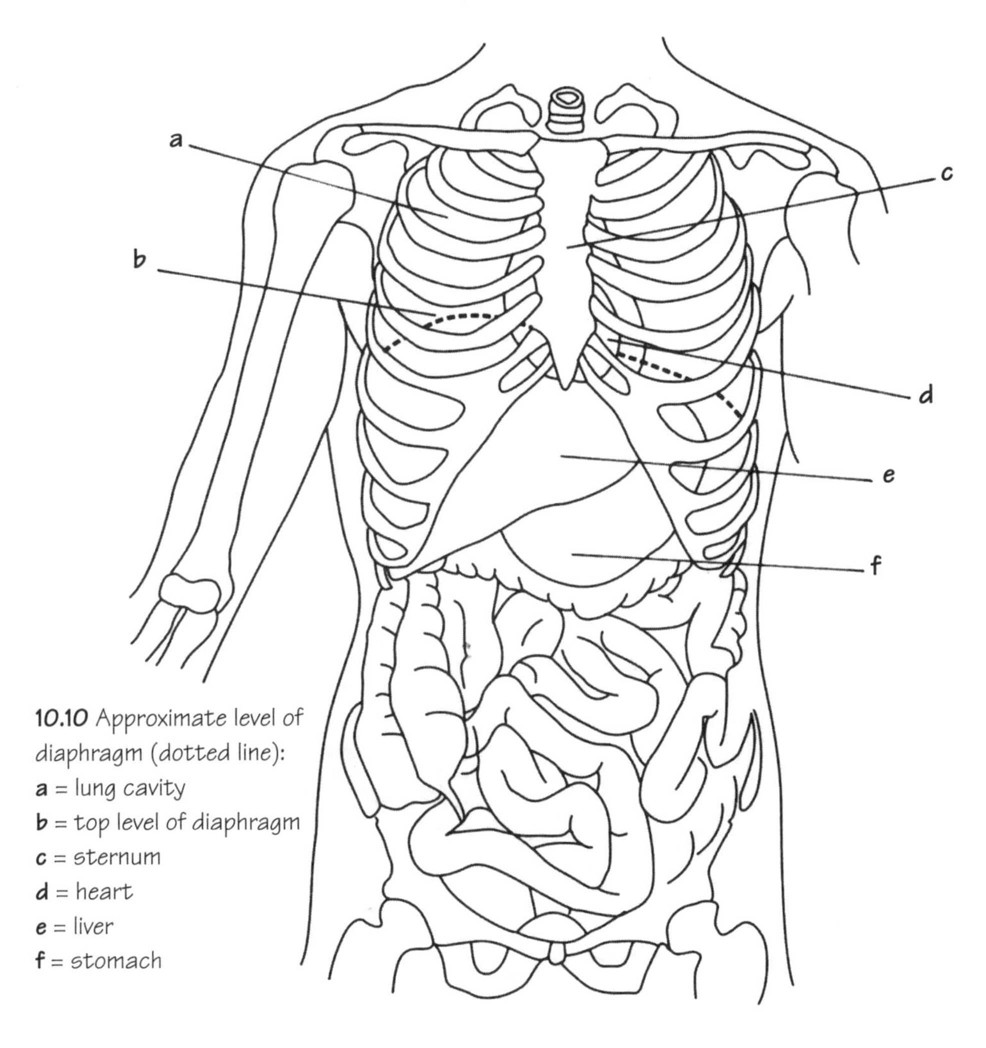

10.10 Approximate level of diaphragm (dotted line):
a = lung cavity
b = top level of diaphragm
c = sternum
d = heart
e = liver
f = stomach

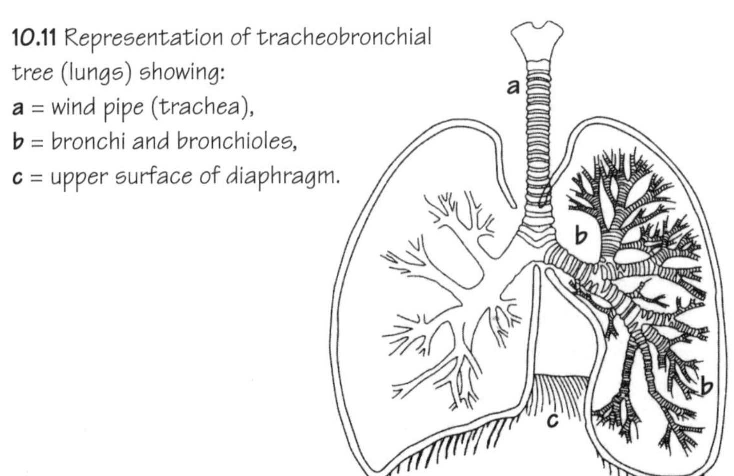

10.11 Representation of tracheobronchial tree (lungs) showing:
a = wind pipe (trachea),
b = bronchi and bronchioles,
c = upper surface of diaphragm.

The diaphragm is often a source of confusion. This may be the result either of not understanding what it does, or not knowing exactly where it is! (See Figures 10.10 and 10.11) Attached laterally to the lower ribs, and frontally to the bottom of the sternum, it then curves, dome-shaped, to the level of about the fourth rib (higher perhaps than many people imagine) and reaches back down again to its lowest attachment inside the lumbar vertebrae. The lungs lie correspondingly high in the chest above it, while the organs of the digestive system are housed below it (see Figure 10.11). The diaphragm is a large, strong muscle, but its sectional construction makes it extremely flexible and sensitive. All we really need to know about it for the purposes of physical training is that it is the main breathing *in* muscle, and because it works *by reflex* it needs no manipulation.

In breathing out we establish strong connections between the body and the throat, paving the way for strong and immediate inspiration. We then discover that two lungs full of air in a body that's in good postural shape are as much as we need for the longest phrase we have to sing, and that instant recuperation (so necessary, for example, in arias by Bach) presents no problem.

Ron Murdock, a specialist in both Alexander Technique and voice training, says that it is important to realise 'just how much the diaphragm is suspended, supported, from above by ligaments which are attached from the top of the diaphragm upward to the spine and front of the chest'. He goes on, 'it is obvious that if one is standing straight and lengthening the spine then the dome of the diaphragm will be raised to its proper position in the body and will therefore be free to work'. [10.3]

4. Posture

Crucial to the full functioning and freedom of the singing voice, posture very often presents a problem in its own right. In fact, good (extensive) breathing and good posture are inextricable. The way we let our bodies go in everyday life often limits our capacity to breathe. If, for lack of exertion or vigour (or singing) we rarely need to breathe deeply, postural muscles feel no need to pull their weight. The normal everyday interplay between those muscles that constitute our breathing system per se and those which provide its structural frame is not sufficiently dynamic to fully satisfy either sphere. This can produce a vicious circle. One example suffices: if we slump when sitting, our ribcage collapses or gets compressed and we cannot breathe extensively. If we don't breathe extensively our postural muscles give up on us. Without this supporting framework we cannot breathe deeply. We may not be able to adequately right this unhealthy state of affairs without asking why we slump or why we breathe shallowly in the first place. If on the other hand our posture is good, it provides a structure already disposed to bracing, so to speak, against any strong movements made in the breathing department. Strong breathing movements, appropriately handled, can reinvigorate this postural frame.

In singing, the correctly extended spine acts like a coat stand for both the head and the torso. Only when this hanger is fully alert and alive can the larynx, torso and diaphragm assume their natural position in relation to each other, and work with mutual benefit. It's only when this is not the case that these organs seem to need special manipulation. Freedom to sing – freedom in other words for the larynx to operate without compensatory help, and for the breathing muscles to operate without manipulation – depends in the first place on the full natural alignment of this structure.

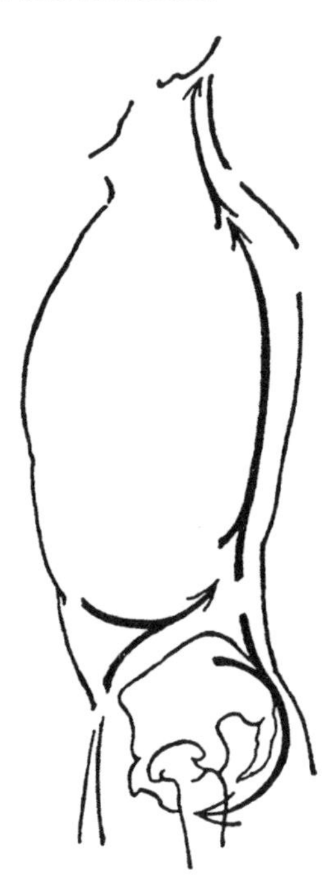

10.12 Diagram indicating body-stretching movements which support the body's breathing apparatus and determine the position of larynx and thorax.

(After Husler and Rodd-Marling)

The relative positions of the thorax and head depend on the spine in its entire length and on the numerous muscles in the back which stretch it upwards (in particular M. *sacrospinalis*, see Figure 10.7) and ground it in the lower abdomen and pelvis. This in turn is affected by how we habitually use or abuse our legs and our feet, which means that, indirectly at least, we sing with our feet, knees, thighs and pelvis, as well as our abdomen, chest, throat and head!

It hardly needs repeating that the four zones described above are not only interrelated but interdependent in the act of singing. Between them they indicate the common major 'fault-lines' in an instrument whose parts are used for many purposes other than singing. The most obvious example is probably that between the breathing apparatus and the larynx, a relationship which is often far from friendly. In our work we must recognise the extent of the splits between the four main zones and evolve ways of satisfactorily reuniting them.

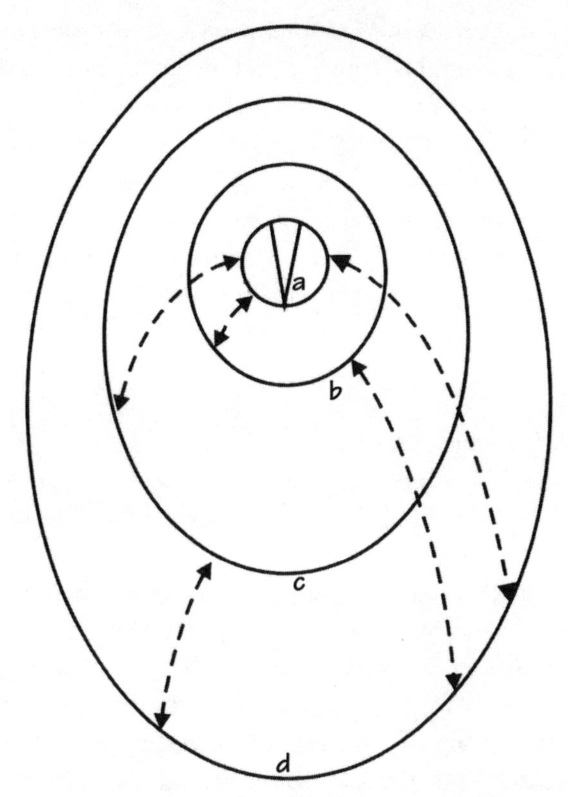

10.13 Audible connections must be made between the various zones of activity: **a** = larynx, **b** = suspensory mechanism, **c** = breathing apparatus, **d** = postural frame.

Before this can satisfactorily be done we must learn to hear. Clearly this is a completely different discipline from either visual observation or theoretical analysis. First we cannot see the totality of the voice in action (only perceive superficially some indirect manifestations of good or bad practice). Second, we can think as much as we like and be no wiser as to either diagnoses or solutions. Singing is one vast coordinated movement. The one movement must be helped back into being by releasing, weighing and balancing the many smaller movements of which it is composed.

PART II

CHAPTER II

Hearing Our Way

It would be wrong to think that the ear is the only useful tool for working with the voice, but worse not to make it our primary diagnostician, constant guide and ultimate authority in the normal course of training. The voice exists primarily as sound heard by both the singer and those listening. The only difference is in perspective. For the teacher or audience the sound is received externally, for the singer internally ('primarily', before the singer has time to 'put it together'). For those who are not used to an aurally based analytical process, the prospect of relying on their ears may seem daunting if not impossible. Indeed, it is fraught with dangers, typically arising from a lack of objectivity or an over-eagerness for results.

When as teachers we first meet a voice we have to assess its strengths and weaknesses, which may or may not be immediately obvious. In any sung sound we can learn to detect aurally exactly what produces it, what's working and what is not, what is under-working and what is over-working. This isn't important either for an audience or for the singer; at least to begin with, when such analysis can be counter-productive. But it is of prime importance to a teacher of singing who, without being able to hear how the sound is being produced, cannot help the singer to correct it.

Outlines of an aural picture

Throughout the whole vocal apparatus, movements and their combinations are reflected as an 'aural picture', providing a teacher with necessary information and the singer, in due course, with self knowledge. Logically, the clearest sound evidence of a voice's working order is to be discovered at its source, at the larynx, where all the movements involved become sound. Ironically, it's from this somewhat divided realm that our ears get their first useful clues about how to proceed towards wholeness. The outer registers (tensing and stretching) provide the greatest sound contrasts. We could begin by describing these sounds simply as hard and soft, because of the relative absence or presence of air in them. Since every voice is different, however, and every singer may have his own idea of what sounds hard or soft (or rich or warm, thin or round) a teacher must appeal to the singer's impression of his own sound if such descriptions are to be aurally useful. Whatever the teacher's and pupil's perception, the important thing to determine in each individual case is how well the folds stretch and tense, and to what extent these activities support or are at odds with each another.

Put simply, we have three main sounds: that described variously as 'head voice' or *falsetto*, 'chest voice' or 'chest register' and a mixture of the two, which we call 'mixed voice'.

Mixed voice

Since the extremes of vocal sound are most discernible at the extremes of a person's vocal range it is clear that the teacher and pupil have no choice but to work at and from these opposite poles, first to clarify and strengthen the sounds in themselves and then to mix them together. In other words, we must work from where the stretching and tensing of the cords are most easily attained towards where they're not – from top to bottom and from bottom to top.

Here is a potential 'stumbling block' which appears to be a difficulty in perception for singers. Voices don't go 'up and down'. The vocal cords lie more or less *horizontally*, and this is the plane in which they make their pitch adjustments. If the voice (its structure) is properly 'mixed' and adequately strong we don't have to reach for high notes or dig for low ones. People also confuse sound or timbre with pitch. This is unfortunate as it perpetuates the myth that chest voice is for low notes and head voice for high ones.

The sung tone is dependent for its technical prowess (at any pitch) on the integration of its separate mechanisms. When this integration is fully accomplished, and only then, 'high' and 'low' notes will lie in the same plane and be felt and heard to do so. Every singer is born with his or her own potential pitch range. So long as we entertain the idea of registers being 'designed' for different pitches, we're in danger not only of skills remaining at odds with each other, but of being misled into thinking that the tessitura (mean pitch) of a voice lies higher (strong head voice) or lower (strong chest voice) than it actually does as a natural whole. However pleasing they may sound, 'good low notes' in a voice may not be as good as they seem in terms of the whole if they bear no relation to the voice's high notes, and vice versa. The larynx is constructed not only for variable pitch but for variable timbres and dynamics at *any* pitch. Singers and teachers must constantly aim for *structural integration.*

Integration of registers is never a question of force. Pushing the chest voice higher and higher will only worsen the divide between this register and the others. Our brief must be to clarify and strengthen the various vocal processes – stretching, tensing, opening and closing – all the while 'testing' or measuring them up

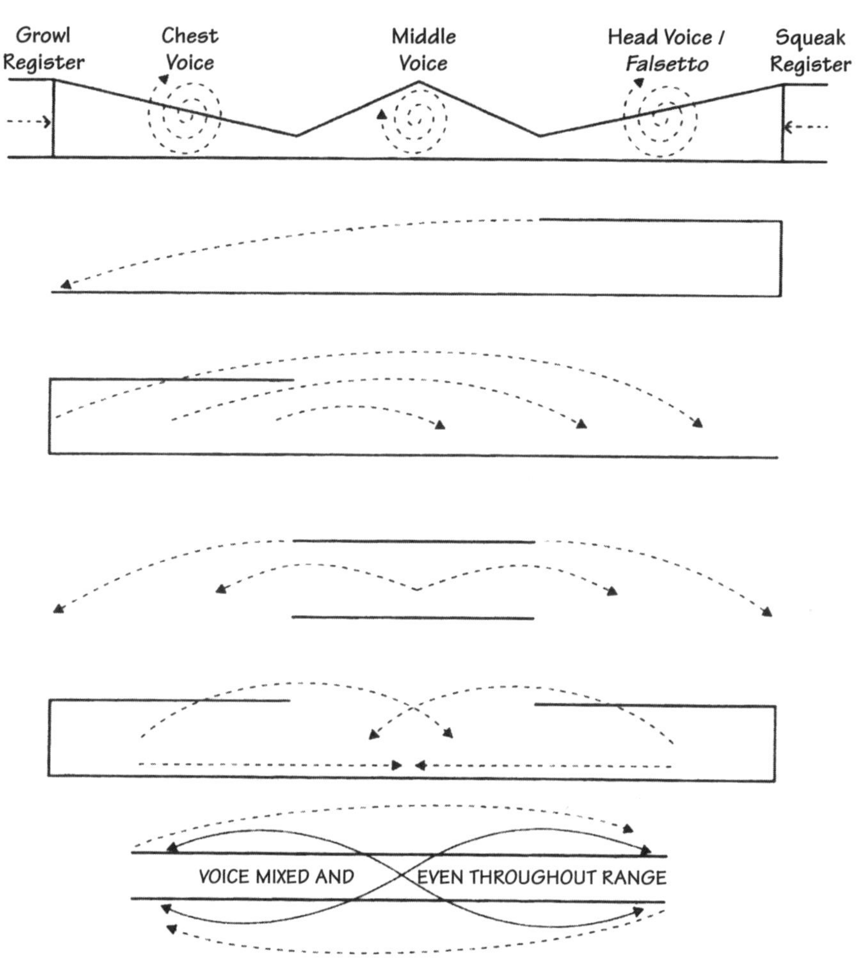

Diagrammatic Representation (only) of 'register-mixing' processes. Growl and squeak registers are sometimes useful approaches to chest register and *falsetto* respectively.

against each other. To this end, for example, we could stimulate the chest voice on appropriately 'low' pitches, then quickly vocalise at a suitably 'high' pitch with the *sound* of the chest voice still in our ears, and its action still in the 'memory' of the muscles concerned.

If a satisfactory mixed or whole tone is to be achieved throughout the range, the stretching processes (*falsetto* and head voice) must be strengthened right into the range normally occupied by the chest voice. When this is achieved, the amount of tension required to produce the 'mixed' sound is minimal. Thereafter, the amount of tension that can be effectively added without strain must remain in proportion to the strength of the stretching. *Falsetto* is often weak to begin with and easily disrupted, especially in cases where chest voice dominates (often the male case).

As we integrate head-*falsetto* and chest voices to form the *voix mixte* we notice a weakening of the physical evidence of a register as it is led towards the limits of its accustomed range. The imperative to strengthen each register in its own right is clearly indicated. Remember that registers are merely parts of a disunited whole. They want to integrate and will do so given appropriate attention and sufficient time. Note also that the 'position' of breaks between different vocal zones can vary from one voice to another. They are normally but not necessarily flanked by weak pitches from the neighbouring zone.

Gender divide

Sexism down the ages may well have helped to perpetuate the 'acceptance' of vocal registers. *Falsetto* is often perceived as an unmanly sound, while chest voice is construed as a mark of virility. The irony here is that the maximum effect and real power of vocal fold tension (chest voice) is achieved only by marrying it with its opposite, the stretching process (head voice, *falsetto*). An over-burdening of vocal fold tension restricts vocal freedom and thereby the full voice of the (male) gender. Similarly, in the case of a female voice, 'shying away from' the chest voice and its integration denies the voice its full power and 'say'.

The missing link

What finally reconciles these apparently disparate vocal elements and makes sense of them anatomically, tonally, and, I believe, psychologically, is the so-called middle register. The sound of this crucial part of the structure is difficult to describe. It is rarely heard, except in a natural voice, in which it is so well integrated that, but for results such as the ability to sing a beautifully projected *piano*, and to execute a finely graded *crescendo* or *diminuendo* without effort (the traditional limited meaning of *messa di voce*), you might hardly notice it. It is like a 'mini version' of the chest voice, but not nearly so heavy or thick. It also resembles *falsetto* with the tiniest amount of vocal fold tension incorporated – a fine, flexible, concentrated sound, which is truest in reflection of both the singer's personality and her intellect. This crystal clear vocal 'core' ultimately defines the individual voice, enabling its 'message' to come over 'loud and clear' even at the voice's quietest moments.

Beginning our aural quest

The neutral throat

The starting point for progressive vocalisation at any stage is a throat without tension. As in singing itself, the throat is induced to work not by physical coercion with mental instructions, but through spontaneity and imagination. However, our throats tend to be tense much of the time through the stresses of everyday life, and even tense themselves in 'readiness' to sing. In order to proceed in training we must first rid the throat of all tension. This neutralising can be done as often as required in the course of a training session, and is a simple enough matter if treated as such.

Experiment 11.1

Lightly breathe (instead of voicing) a neutral vowel, preceded by 'H', as in 'HER'. The jaw and tongue should be slack. What happens? So little that perhaps you hardly notice it. Be observant, but don't analyse! Have another go. Already you have the essential ingredients of a panting exercise in which air is gently exhaled – without control or restraint – and the body has no choice but to follow it. Nothing is pushed and the ribs respond as if by collapsing. *Body movement is induced* by what is (or is not) happening in the throat.

Experiment 11.2

Purse your lips into a loose 'F' shape and blow as if to send a constant stream of air across the room. Note that this tends to engage the body more actively than **11.1**. But notice from where the air begins to move – from the lips – or if you do this with your mouth closed, at the nostrils. If you observe this, you will be less inclined to manipulate the breathing system.

Go back to **11.1** and lightly but vigorously 'huff' a quick, short rhythmical pattern:

Panting begins! But only if you *let* your body respond to the out-huffs of breath. If you find it difficult to let your belly go, try panting on all fours like a dog! The belly should hang free so that, at the end of each rhythmic movement, there's nothing to let go of.

Note that when the panting is freely done there's no need to take breath; the organs lying below the diaphragm are left alone and move only *by influence* of the thoracic movements induced by the panting. In this simple breathing action the breath returns to the lungs by force of gravity, or the weight of the slack bag of organs! You might get a better idea of this effect by sitting with a straight back, arms folded across your belly. This simple movement is frequently frustrated by the Western cultural habit of holding in the tummy.

How happy the throat is to remain in this neutral state may be tested by continuing to pant while loosely and arbitrarily changing the vowel shapes and occasionally closing the mouth altogether. With practice none of these changes, or changes in speed or strength of movement need produce tension in the throat or manipulation in breathing. 'H' is the softest consonant, but often gets stuck in the throat (as in the Spanish '*jota*' or the guttural sound you get when the tongue and the soft palate are nearly touching). Try a strongly whispered 'Happy' and you may hear a slight rasping in the throat which indicates that the passage of air is not entirely free. You might also observe that the body 'kicks in' in response to a throat under tension, but follows without bidding when the throat is free. Throaty-sounding tensions (however produced) can sometimes be eliminated by imagining that the exit of breath is on a level with your sternum, or that the windpipe ends there, or has a much wider opening than it does!

Playing with sounds can be most instructive to the ear. The tense and neutral throats provide us with two extremely useful and quite audible contrasting sounds, although, strictly speaking, there's only one neutral throat and an almost infinite variety of tense ones. In addition, we can hear the physical *extent* of the sounds (whether deeper or shallower), and whether this is manipulated or naturally achieved. We will see that the difference between a neutral and a tense throat is highly significant when it comes to voicing (as distinct from speaking or whispering) vowels. We can learn to differentiate between unhealthy tension, blocking our way forward, and tension (whether it is the result of intrinsic or extrinsic laryngeal musculature) which indicates healthy voicing.

Unhealthy throat tension is that which anticipates movement or restricts its freedom, whether at the larynx or far forward of that, where speech demands its own freedom of articulation. The experiments above have begun to give us an aural inkling of the influence that the throat has on the body.

Neutral throat and vocalising

As throats are rarely free of tension, panting can be developed and used as a throat-neutralising and body-relaxing exercise, and returned to as often as necessary for short bursts at a time. However, panting does little more than improve upon our everyday breathing and pave the way for something more positive. If neutralising the throat is to be of any real use to the singing voice it must be strongly aligned to *breathing out* and its reflex of breathing in. At best, panting encourages an unobstructed, unmanipulated flow between torso and throat, the breath seemingly a mere by-product, not dissimilar from normal 'quiet' breathing. For the relationship between the throat and the thorax to really come alive in the cause of singing we must invigorate and strengthen this reciprocal relationship.

In various workshops we have had fun blowing out candles. When people blow out candles (even a few on their birthday cakes) they take a big breath, anticipating the need for strength and duration of blast. What in fact matters is not air quantity but its direction and concentration. When conducting this experiment there are three rules: 1. Relax your belly and don't deliberately breathe in 2. You have one puff per candle 3. Get on with it!

Participants file past a candle which is instantly re-lit, and sometimes moved further away to improve the performance of both adepts and those breaking rule one!

Experiment 11.3

For training, it would be better to set up a row of candles (five or more) and for one person to puff out all five in quick succession. Breaths must not be deliberately taken but the puffs must be completely new for each flame. The interesting thing is that, done well, even with many puffs, there is still no need to take a breath; the whole operation is done spontaneously, and the breathing apparatus resumes its relaxed/alert shape – automatically – the instant the puff has been executed, unless you have held onto it or tried to force the issue!

Developed, this short, sharp exhalation will have a profound effect upon the whole vocal system, since it eventually draws upon strong contributions from the spine-lengthening muscles as well as those suspending and supporting the larynx. Vigorously carried out, this extensive but naturally-coordinated movement serves to remind the torso of the totality of its contribution to the singing voice while preparing the throat to be in a better condition for responding unimpeded to our expressive intentions. Providing the throat is neutral and we leave our breathing system alone, it will respond suitably if at first weakly to any clearly imagined, currently attainable sound. The better shape the breathing system is in, the more readily it will respond without fuss. Here we have a guiding principle which will serve us throughout the training process: **Aural imagination calls forth a sound which induces a corresponding movement** (the one that goes with it) **from the body.**

Vocal sounds are not absolutes, and nothing is to be gained by treating them as though they are. It is only when they are *released* that they are given identity. Strange as it may seem, once we recognise a sound, we may no longer feel the need to describe it. If on the other hand a teacher says to a pupil 'Yes, *that*! That's chest voice', he runs the risk of labelling an 'effect'. If a pupil says he's 'got it' it usually means he has *felt* rather than *heard* it. Next time 'chest voice' is on the agenda this pupil is likely to look for the sensation rather than listen for it.

Portamento

The *portamento* (something heard rather than felt) can be physically freeing, aurally connecting and informing and emotionally evocative. This sliding motion is a quality of many exclamations, such as sighing. In the voice, musical notes are not fixed entities; sliding or stretching is essentially how the sung tone moves from heard note to note while sustaining itself intact. *Portamento* is also the physical basis of *legato* and therefore needs to be cultivated (even indulged in to begin with). When two sung notes are joined, the sliding between them is so subtle that it is barely noticed, but we must be aware of how notes are elided to ensure that individually they don't interrupt the flow and phrasing of the sung tone.

Our primitive ear recognises vocal tone as continuous; this is the basis of vocal lyrical expression. As our voice becomes more finely integrated, we can increasingly depend on this 'unthinking' ear to monitor the tone's course. There is no musical instrument that has such a range of notes and colours embodied (literally) in such a small mechanism as the voice. We need to understand what this means so as not to make disproportionate physical movements and vocal gestures. The *portamento* helps our ear to coordinate and sustain the voice's minute activity. With its aid we can not only target specific notes with pin-point accuracy but detect just how present or elastic any element of vocal *legato* is. In training, the singer has only to follow a sound by this means to observe how far it will go intact and to monitor the slightest variations, keeping it on track or encouraging it to go further without neo-cortical intervention.

To breathe or not to breathe

Arriving at 'useful' sounds in training is assisted by the singer's developing aural imagination as the teacher guides the 'sound play'. In genuinely *imagining* sounds, rather than thinking about them or feeling for them, the singer is invigorating the natural connections between her ear and her voice rather than giving way to breathing manoeuvres which simply don't tie up with what's going on in the throat. This use of imagination gives both singer and teacher the freedom to explore sounds which neither is likely to have known were there. It avoids the struggle to make the voice fit preconceived notions of 'how it should be' and makes way for the discovery of something natural and personal. On another level, it gives the pupil freedom to experiment and 'get it wrong'.

To further the argument, we can say that if we close the cords, either by means of vocal fold tension or by the closing mechanism (exemplified and partially isolated in the *coup de glotte*), there is bound to be less air in the sound we make. If we open the cords or stretch them (without the tensing or closing) there will be substantially more air in the sound. The *quantity* of breath is a logical outcome, not something we choose, and can therefore be considered as incidental. In neither the closed nor open example is there *excess* air – the sound of opened cords isn't 'windy' and that of the closed glottis isn't 'pressed'. If a voice is hard or tight sounding, stretching and opening exercises are clearly indicated, rather than the need for more breath. Such sounds will sound *relatively* breathy, but a distinction must be made between breath *content* in the sound, resulting from active stretching or opening, and *breathiness* (excess air) which is the sound of weakness or flaccidity at the glottis. Conversely, a sound which *is* breathy indicates that stimulation of the tensors or closers is needed – not holding breath back or putting the breath under pressure, which causes unwanted tension or resistance in the throat.

Energy, vibration and movement

The energy that feeds the sound and the sound's vibration are as aurally observable as movement from note to note. The two elements should not be confused; energy is required to bring a sound to life, while the vibration of a sound *is* its life. By movement in this context I mean vocal gesture, in, for example, a musical interval or phrase. The sung tone as distinct from the musical phrase is a composite of characteristic vibrations. By means of a *portamento* we can easily tell whether a 'head-voiced' 'Oooh!', for example, is freely delivered (freely vibrating in accord with its character), in which case it will remain elastic and seem to move of its own volition, or whether we are attempting to manoeuvre it over a series of notes in a relatively fixed fashion. If we have inhibited vibration or physically interfered with the movement the sound will travel little distance in vocal range without some deliberate change of emission (usually another form of fixing) or it will collapse.

Falsetto and chest voice both have distinctive vibrational and therefore tonal characteristics (the one the disembodied sound of the stretching membrane, the other the 'meaty' sound of contracting vocal folds). If we mix these processes, and if the tonal characteristics of both are heard in combination, we have successfully rolled two throat-body movements into one. We can balance opposing forces simply by listening, but satisfactory mixing will only occur if the separate movements are sufficiently flexible (freely vibrating) in themselves. A rigid movement will not gel satisfactorily with a freely vibrating one any more than it will with another rigid movement. Movements which merge satisfactorily can be heard to have done so almost as easily as others can be heard to be at odds with each other.

For the purpose of visualising this idea of 'movement within movement', I was once told that Frederick Husler used the analogy of a pocket watch in which many cogs make small but indispensable contributions to the efficient performance of the whole. While the 'hands' move smoothly round (let's call this the musical phrase), this is the outcome of the interdependent cogs working synchronistically. The 'co-operation' of movements secures the smooth running of the watch (sung tone). Stop a cog, and at the very least the mechanism will begin to fail. Similarly, our *legato* will be less fluid (linearly and laterally) if a strand of vocal movement falters or breaks down. We cannot see these vocal 'cogs' but they compose our aural picture, reflecting any variations in form or fluidity. Once it is clear, the singer's aural picture enables him to both initiate healthy tonal fluctuations and correct unhealthy ones while the tone is in flight.

Sounding emotion

We can also hear the emotional content of vocal sound. This could be called 'tone of voice'. We all know the difference in effect the same words can produce depending on the tone in which they are spoken. The tone we use to say 'Stop!' or 'Jump!' could make the difference between life and death. 'I love (*love)* you!' or 'I hate (*hate)* you!' could illicit quite different responses or actions from the person addressed. The potential of 'tone of voice' becomes more clear, and its realisation more inevitable, as a singing voice develops because it shares the roots and the physiology of emotional utterance, our primary vocal motivation. The employment of a *portamento* can help us to tap into and integrate the emotional content in the singing sound by playing with and encouraging spontaneity of shape and inflection, the essence of musical expression.

While in the course of training we can and must look for these three naturally interdependent components – tone colour (characteristic vibration), movement (musical inflection), and emotion (comprised of both these) – we can also use them as 'releasing tools' in their own right.

Vowels

In order to amalgamate the various laryngeal processes, it's important to know which vowels lend themselves to which processes. This also helps in our aural imaginings and in clarifying the relation between words and the sung tone with its myriad colours. The idea of 'vowel equalisation' should be understood not as an aim in itself so much as a means of achieving vocal integration, since vowels in themselves are not a problem.

In recognising the allegiance of vowels to vocal processes we can easily detect in a phrase of words and music where weak points in the *legato* occur and why. For example, the well known difficulty in concentrating the sound on 'oo' is explained by the fact that 'oo' lends itself to the stretching and opening rather than the tensing and closing processes. Thus 'moon' is likely to sound more breathy than 'wax'. 'Wax' on the other hand might sound inappropriately sharp because its vowel sound lends itself readily to the tensing of the folds.

In working with vowels, we can learn to hear and understand more clearly what has led to the non-singing quality of the speaking voice. The vocal cords give everything away. Generally in the normal speaking voice we hear wide variations in breathiness and concentration between vowels because their formation influences the shape of the glottis and how open or closed it is. The formation of vowels varies not only from language to language but from speaker to speaker of the same language, adding further to the fascination as well as the individuality of our work.

Singing requires that the tone (its formation) is constant, if only to provide cohesion and maintain the relationship between all the other variables which are specific to this activity – dynamics, colour, emotion, freedom to articulate words, projection – to make integrated sense of the combination of attributes that we call singing.

Working appropriately with vowels can restore balance and stability to vocal tone so that changes of vowels in the course of singing don't disrupt it. If there is confusion between what produces tone and what produces words, and if these separate departments are not free, attempts to equalise vowels can easily degenerate into a vowel modification in which all vowels sound roughly the same. Modification of vowels can lead in other words to structural distortion. As the formation of a singer's tone becomes constant it is increasingly easy for her to experience just how slight movements of the tongue and lips need to be and how deft they can be when articulating words.

For centuries, the pure Italian vowels have been recognised as the most suited to coaxing the voice back to its singing condition. This is because in their purest form the primary vowels are the ones most closely aligned to the basic vocal processes. Thus, for example, in addition to 'u' or 'oo' already mentioned in conjunction with stretching and opening the folds, 'ah' is best for stimulating their contraction, and 'ee' can help to stretch the vocal bands (mucous membrane) – *falsetto* – and (along with 'ah') stimulate the 'closers' as in the *coup de glotte*. Primary vowels are like primary colours – interestingly there are three of each. Working with vowels, we have a tool of great flexibility, which lends itself to the training of speakers of all languages, the singing of different languages, and to a high degree of fine tuning. Pure vowels and by implication fully formed vocal processes may have to be approached gradually via the vowels in the language native to the singer. Nevertheless, the structural health of the larynx can to a large extent be verified by its ability to form these pure vowels fully and freely. Ultimately, it is the natural integration of the acoustic properties (resonance) of the pure primary vowels that brings about the indivdual's ideal singing sound.

Placing

Closely allied to aural imagination and useful in conjunction with suitable vowels is the idea of 'placing' the voice. The idea behind 'placing' is to imagine the voice 'placed' in a specific locality, where, if successful, vibration may be felt. However, the various localities around the head and throat are not like magic buttons producing instant predictable results on pressing! How successful placing has been can be best judged by the sound that ensues, since it is the vocal activity that produces the vibration. We are not groping for a physical sensation so much as using our imaginative and aural faculties to *avoid* doing so. The word 'placing' may be the stumbling block for those who shy away from this approach. Placing does *not* mean locating or positioning the voice or 'putting it' somewhere. It is merely an aid to inducing audible activity. Anyone who has worked in this way with an aurally perceptive teacher will have experienced just how *facilitating* this tool can be, and it's precisely this quality which makes it both effective and healthy. Imagine the discomfort and confusion if all the movements of the larynx, intrinsic and extrinsic, had to be physically felt! Correctly handled and aurally monitored, placing provides an accurate focus for 'aural targets' and enables this to be done with strength and without discomfort.

Progress with any exercise depends on appropriate preparation, and on how the exercise is carried out. Placing may not be effective at first, either because the singer's imagination doesn't lend itself to the idea, because it is misapplied, or the ground is not properly prepared. Successful placing depends in particular on a neutral throat, one that is receptive to 'the idea'. As in singing itself, a voice must be 'available' (not fixed) if the imagination is to be satisfied. Discovering vibrations in various locations is far from a 'technical fix'. There's nothing we can do about *where* the voice is, but we do need a means of animating it without imposing our will on it, without being physically intrusive. So, the objective of placing is not the vibrations in themselves (although these *can* be a useful temporary guide) but the effect it has on the intrinsic operation of the larynx and its positioning in the throat. It's precisely because one cannot deliberately or easily make such manoeuvres without fixing the throat that the indirect method of 'placing' is indispensable.

Placing successfully 'at the top of the head', for example, will have the effect of inspanning the larynx (up and back, and forward and down) while indirectly stretching the vocal folds. When this sound is good we know it as 'pure head voice'. Freely executed, vibrations will be felt high in the head. This is hardly surprising when you realise that this is actually where the muscles concerned are 'rooted', their 'origin'. They will also be felt at the top of the sternum, where the anterior depressors of the larynx originate. As already mentioned, the latter work by reflex when the elevation just described is achieved correctly. To look for this two-way inspanning by placing the voice at the top of the sternum would therefore be trying to work upside-down, a case of seeking the effect without understanding the cause.

This way of utilising the imagination in training needs careful handling, and judicious use of vowels and consonants. Also, we must differentiate between this work and the act of singing itself, when we no more *place* the voice in all the localities (so that all the correct laryngeal activities will spring concertedly to life) than we try to *feel* the hundreds of physical sensations that combine to produce the singing voice. This being said, while a voice is developing, certain vocal processes lagging behind may need to be consciously encouraged – the equivalent in sound of highlighting – while the student is singing. This should be done as an aid to judging vocal balance (like aurally aiming a little more to the right or left) rather than a failsafe procedure. By this stage of progress a singer will be aware that placing while singing is a matter of inducing or stimulating vocal activity through sound or tone colour to stimulate or bring back into focus a part of the aural picture that still tends to be weak.

When the voice is in good balance and strong but flexible in all respects, and once the ear is fully cognisant of its role, tonal adjustment becomes 'aurally automatic'. It is not, of course, always so easy. Even a healthy singing voice will meet seemingly insurmountable problems in vocally unsympathetic music. Occasionally a singer might want to exaggerate tonal bias for the sake of emphasis or effect. In both these cases, deliberate placing might be helpful as a temporary measure.

While the idea of 'placing the voice' was never intended to explain how the voice works, it can be an invaluable way of temporarily by-passing explanations which are often worse than useless to a singer. Many new pupils have arrived with a list of instructions which are thoroughly inhibiting. Beware of placing used as a technique for singing per se: there is no one area in which to place the singing voice, and nowhere to put it. The singing voice can be woken up, animated and liberated, and to this end we need all the suitable tools at our disposal.[11.1]

Mimicry

Human beings are marvellous mimics, perhaps more so with our voices than our bodies. Children love imitating sounds and gestures, and mimicry is considered great entertainment. Imitation is a natural means of learning. Our ears lead the way.

Example and imitation can be invaluable in the process of learning to sing: initiating the process of sound exploration, conveying what the teacher has or has not in mind, and eventually clarifying an individual's true sound by means of comparison and appropriate feedback. However, we don't want the end result of training to be merely imitation, and teachers should be aware of the dangers of copying. Imitation can be a barrier to self-awareness and imagination. One pupil of mine once said of another, 'he sounds just like you' (meaning *me,* his teacher!). While this was meant as a compliment I was duly taken aback, and remember replying to the effect that he wasn't *supposed* to sound like me but like *himself.* From then on I kept a look-out for potential Harrison clones. Many singers do in fact sound remarkably like their teachers, invariably to the cost of their vocal freedom and true individual identity. A singing teacher should be aware of this possibility, and so should pupils.

To whatever purpose it's put, imitation demonstrates the astonishing versatility of the human vocal organ. A child learns to speak by imitating voices. As he does so, the muscles of his larynx, particularly the intrinsic ones, become 'programmed' in two ways. First, there's the parent's native language, the vocal 'pattern' of which might easily be mapped, since the glottis assumes varying degrees of closure and a different shape for each practised vowel. In other words the larynx is muscularly programmed vowel by vowel.

The second influence on the formation of the speaking voice is more complicated, relating to the personality of the parent(s) imparting speech to the child

– the 'vocal personality' of whoever is responsible for the teaching. On the surface, this could account for measurable vocal defects, such as weak closure of the glottis, through the imprecise articulation of consonants, the exaggerated high pitching of a voice which by nature lies lower, or the deepening of one which lies higher. On another level, embedded in the teacher of speech's vocal programming are the physical manifestations of his or her individual emotional influences. These result from personal emotional biography, or from emotional patterns 'adopted' from his or her own primary language teacher.

While it would be perverse to pursue this line of hereditary investigation in training, a singing teacher must endeavour to discover what in an individual's sound is true to him or her, noting that the character of a speaking voice owes as much to emotional as it does to physical patterning. The teacher should soon be able to ascertain, for example, whether her pupil's sound is the shy, polite, reluctant or demonstrative version of itself, or whether it 'rings true'.

While the vocal apparatus is capable of producing a great range of so-called singing sounds, it can also make an astonishing variety of sounds which, although still described as vocal, are far removed from the singing voice as such. This versatility is likely to prove a valuable 'tool' in our more radical work. When I demonstrate a 'funny' sound for a pupil I often get him to describe it. So, if it sounds to him like bleating or a foghorn I ask him to produce his own version. This has proved an effective way of circumnavigating feelings of embarrassment or resistance.

Today, different vocal sounds are being formulated and scientifically explained so that they can be made by anyone who has at least some degree of vocal freedom. It can be argued that this is good, because it reduces the possibility of doing damage to the voice. Taking the holistic view, however, this is a highly contentious training departure. Vocal sounds are not abstract entities: even spontaneous 'non-singing' vocal sounds originate from the voice without arrangement. Reductionism applied to the voice limits the singer's legitimate choice of sounds to those that can be categorised, explained and reproduced consciously and 'safely', and seriously affects spontaneity in producing them. Worse still, this formulating process seriously interferes with the natural, reciprocal flow between the aural imagination and the singing voice as a whole.

If all its working parts are fit and responsive, the singing instrument *comes into being on impulse*. Thereafter, the myriad of singing sounds, already different in each and every throat, are harmoniously achieved by means of more or less subtle bias. The proportions of the many elements that constitute what we call mixed voice are varied, not through force of will but through expressive intention, imagination and feelings. There shouldn't be any 'gear changing', fixing, separating off or 'repackaging'. To achieve its potential variety of colours and shades, the vocal instrument needs to have become organic in operation. When the natural functional integrity of a voice fails, it loses not only its natural musical and expressive skill, but what we describe, at least in classical terms, as its 'singing quality'.

Although our learning to speak causes a certain degree of disintegration or degeneration, we are not aware of this going on, not conscious of what we might be losing and of the vocal limitations we might be bringing on ourselves. Our speech voice is the one with which we most closely identify, and therefore we learn unwittingly to identify with our limitations.

The process of rediscovering our original voice and its potential is necessarily more conscious than learning to speak and we need viable ways to proceed without worsening the situation. The singing voice is much more than a specific sound suitable for singing songs. Its deeper roots indicate that in reclaiming this voice we are sounding out both our individuality and our shared humanity; a two-fold expressive identity.

It is interesting to observe that all the great singers that one can think of possess their own individual sound – there are no stereotypes. Trying to be a 'typical' soprano or tenor is trying to produce something which doesn't exist in reality. Perhaps it's because there are so many *imitations* that there are so few great singers. Subjectively, although perhaps unconsciously, the singer who emulates the singing of another is somehow avoiding his responsibility: to himself as an individual, and to his audience whose collective humanity he is representing or reflecting. A singer must find himself in the mirror of his own sound before he can be a mirror for his audience.

PART II

Chapter 12

Trial and Error

Introduction

The teaching of singing is an individual matter. While each teacher must find ways and means which suit his or her personality, these must be sufficiently flexible and adaptable to cater for every individual singer with his or her changing needs. A pre-formulated technique is by definition a denial of individuality. Teachers should be appropriately inventive and experimental in their approach to their work.

The quest to determine scientifically how the voice works may have contributed as much to the standardisation of singing (the general absence of individuality) as the ready availability of vocal recordings, the German *Fach* system and the proliferation of vocal techniques with fixed agendas. In teaching singing we're dealing with the constantly-changing physical-emotional state of a person whose vocal condition has been arrived at via a unique physical-emotional-intellectual-spiritual journey over many years. No one and nothing can properly instruct us how to work with a voice except the unique being to whom it belongs.

It is easy to fault a voice, or even to imagine it fully developed, but the solution is rarely if ever a matter of replacing bad practice with good, which is more likely to lead to further conditioning. While most techniques are designed to gain a preconceived product by some prescribed means, a viable *process* is a constant advancement by unwritten means towards the freedom and completeness of a voice with its own unique dimensions and qualities.

Steps or stages in the restoration of a voice which have been overlooked or given insufficient attention can be difficult if not impossible to return to later. Good training provides a basis for improvement at every stage; something to build upon, which can be monitored and judged better for what it *is* or what the voice can do naturally than against an expected outcome. This chapter explores some enabling devices and procedures.

Our job as teachers is to facilitate the voice's natural attributes by natural (physiological) means, bearing in mind that the emotional state of the singer has a direct impact on his ability to vocalise. Training must be in tune with individual needs and must reflect all the human aspects of which singing is made up.

Creative tension

In singing, the process that may provide us with the most flexibility and motivation is similar to any creative process. Firstly, we must have an ultimate vision or goal. For a singer this will be related to his or her aspirations towards performing. It must be clear, detailed and realistic. The singer must keep one eye on this goal but not fret when it seems a long way away, or a long time in coming nearer!

The second necessity is to know and appreciate the 'material' you are working with. The teacher must help the singer to gain an appreciation of what she already has or *is* in terms of the singing voice. Singer and teacher work together with the same material. Notice this is a question of focusing on what exists – certain strengths, freedoms, tonal or expressive qualities, energy and so on – not what's lacking. The weak or stuck parts of a voice can be approached by way of what is already available: for example, breathing can be used to neutralise the throat, making it 'available' for the stretching of the cords. The stretching in turn paves the way for the introduction of tension. Once released, if only to a small degree, development can proceed by extending each operation or combination of operations gradually over the singer's natural range. The process finds its own way by assimilation, throwing new light on its tasks, invoking fresh ways of thinking and acting, and leading to greater understanding and skill on the part of both singer and teacher.

The antagonism between what you want to become and what you actually are produces a tension, which in the words of Robert Fritz 'strives for resolution'.[12.1] This tension can seem to act in a contrary fashion during the period where we're not 'where we want to be', pulling us forwards and backwards, towards our goal and away from it, depending on our current state and how we're dealing with it. While initial progress might seem quick, we are easily pulled back towards a feeling of not achieving, especially when new difficulties emerge.

Although working with what we have or where we are is logical, progress is unpredictable and we cannot know in advance what's going to happen. The nature of the work varies as the discrepancy between the goal and the present reality decreases. Fritz says 'Because creating is an art, it deals in approximation. There are no formulas to follow, no hard rules to apply. Creating is, at its roots, improvisational. You make it up as you go along. You learn to break ground. You learn to learn from your failures as well as your successes. Over time your own unique creative process develops and your instinct for making choices that are in your own best interest increases'.[12.2] Once the elasticity of this idea is taken on board, it can have a liberating effect for both teacher and pupil.

The spiral of progress

The idea that progress is linear, 'a to z', is clearly incorrect when we consider the voice as something organic. It is tempting to gauge success by the attainment of useful items, such as high notes, loudness or being able to sing for long stretches, but if we consider that all these facets are interdependent, we realise that we pursue one or two of them at the peril of damaging or falling short of the whole. Thus, high notes might be achieved at the expense of flexibility, loudness at the expense of quality, and so on. In a viable process of vocal work we must attend to all the facets of the voice, strong and weak, in proportion. This doesn't mean the mathematical division of time spent on *forte*, *piano*, *legato*, high notes. On the contrary, such 'skills' are intrinsic to the singing voice, and it is the voice as a whole that needs to be released and developed. Work must be apportioned so that the structure of the singing voice will gradually come together, and the innate vocal capacity of the individual be *realised* as a whole. There is no magic in this, only systematic, thorough and meticulous hard work.

A viable process must espouse the idea that 'answers' always seem to be questionable – perhaps because change and development always give birth to new questions. Work changes as it progresses, with shifts of focus, emphasis and direction. The eclectic nature of the singing voice and singing itself is duly catered for. As the singing voice comes back, so does its ability to be employed in the service of words and music, and the ability to satisfy this intellectual activity is constantly checked by pupil and teacher against the *physical* progress. The physical and the intellectual are thus integrated logically and by degrees, avoiding complicated or effortful premature attempts at compatibility.

The image of a spiral indicates development and refinement, achieved by constantly 'doing the rounds' of all facets of the instrument from the deep postural superstructure to the finest margins of the glottis. At each slight turn we see where we are, rather than where we 'should be' in relation to the whole, working from reality, not false assumptions. This also helps to ensure that we lay firm foundations for each stage of progress. The reluctance to move in the sensitive and crucial area of the emotions may call for spirals of careful affirming work. Spiralling work can enable us to check out progress from different angles. Constantly changing the perspective can help us to work thoroughly and incrementally without cutting corners, and without getting stale. The work should always be fresh. Indeed, however well we may think we know a voice, we should always begin work as though we are meeting it for the first time. In this way there's no room for boring routine or mindless repetition. Because it can be difficult for the singer to assess progress, it's important to constantly point out what is being achieved incrementally – a little more strength here, greater ease there – as we proceed. Returning constantly to every facet of the instrument and weighing it up against the whole helps the singer to become self-aware and gain confidence, while assisting the teacher in measuring balance, strength and details.

The spiral suggests opening up or unfolding, allowing for individual progress

, flexibility in all aspects of training, quicker or slower progress as condi- :tate, and dipping back to earlier work if it needs reinforcing, all without ne general direction.

Normally, training begins with a broad or rough base and becomes more specific and refined. Unlike the linear approach, which suggests aiming at a specified goal and getting there as fast as possible, the spiral suggests creating something by working on whatever is discovered to need attention, and taking as much time as is needed to do the job thoroughly.

In accord with the creative process, the spiralling movement inspires changes of direction. In the context of vocal development this means that we can be surprised by a turn that a voice takes in terms of range, quality, ease or emotional connecting, without being thrown. This is because we are not controlling the voice; we are facilitating it. It can suddenly be inspired to take off! Such unexpected moments can be the most pleasurable and exciting of the process.

It would be misleading, however, to suggest that beyond a certain point the singer is necessarily 'home and dry'. Fine tuning can be tricky, especially if the singer is currently attempting repertoire which is unsuitable. Also, we are continually bringing new life experiences to our singing and ability to interpret and communicate. It would be good to think that these experiences will improve or enhance our singing, but this isn't always the case. A more contented or stable person can feel less motivated to improve, and a bad experience can raise a singer's performance to something extraordinary. So long as a singer's objective *is* to improve and she is sufficiently knowledgeable about her own voice (herself) she'll endeavour to assimilate her experiences, professional and personal, so that they contribute positively to her performance. After all, life, whether experienced or merely observed, is what singing is all about.

The following sections explore some devices that can be used to stimulate the creative process of training.

Imagination and perception

Imagination plays a huge role in any creative activity. A child has no difficulty with this, giving his toys personalities and making up stories about them. Practically anything has imaginative potential for the child. In our current society, however, so much is made obvious or taken for granted that our imaginations can become lazy. We may even have become afraid of imagination. In his probing and perceptive study *An Intimate History of Humanity* Theodore Zeldin writes: 'Imagination has long been looked upon as dangerous. The Bible condemns it as evil (Genesis 6:5) because it implies disobedience. Even those who have wanted to liberate humanity from tyranny have feared imagination as a threat to reason. The philosopher John Locke (1632–1704), who was generally an enemy of all dogmatism, warned parents who discovered a "fanciful vein" in their children to "stifle and suppress it as much as may be"'.[12.3] In training, however, breaking the rules of habitual life may be necessary to release ourselves from limiting or negative conditioning and break new ground in our development.

Peoples' imaginations can be vivid and detailed as well as fantastic. In the performing arts we are concerned with an inner journey in which imagination can become our reality. Imagination enables us to see things in a new light, giving us a chance to reconsider what hitherto we may have blindly accepted as correct. Imagination can show us other possibilities, if we dare to look at them. In singing, a major difference between a competent professional and a true artist is the element of creative imagination that the latter explores to memorable and inspirational advantage. It can be scary to be overtly different, to be innovative, to explore uncharted territory, and society is often wary (or envious) of the individual. However, the term 'individual' can mean 'indivisible' (not divided against oneself), and insofar as we are unique, I believe we owe it to ourselves to discover what that means so that we can enjoy being fully alive.

Imagination in our studio work is of vital importance in problem solving, which is not so much a matter of putting things right as of bringing things into being, or re-creating. Changes don't have to be dramatic and simple devices can have significant results. For example, some people find it beneficial to imagine tone in sound as tone in colour. The major task of integrating the registers, for example, can be made easier in this way. Fun can be had finding just the right shade in which *falsetto* (usually perceived as a light colour) is 'holding its own' or is not blotted out by too much darkness from the chest register. Imagining in colour is quite apt in this work as it is more akin to hearing than to physical sensation.

Tone sense

Many singers seem to lack what I call 'tone sense'. I have come across a number of would-be singers who seem to have no sense or appreciation of the tone quality they are producing. Not surprisingly, this has been more prevalent amongst people who have tried to 'sing on their speaking voice' than those with an idea about how singing should sound. This problem is too serious to brush over as it isn't just in more subtle work that singers fail to recognise the qualities of the different sounds in themselves. This recognition is as important to a singer as recognising shades of colour is to a painter. Even relatively obvious tonal differences, for example between head voice and *falsetto*, or chest voice and closers, can be problematic to begin with. Teachers must constantly encourage singers to notice and describe the subtlest tonal differences until such discrimination is second nature.

Tone sense is related to quality, something difficult to describe without suggesting that the aim of singing is simply to make a beautiful sound. Every singer's tone quality is of course unique, but to discern *true* quality (as distinct from that which is not 'true' but is perceived to be good or beautiful) can take diligence. Singers and teachers should be encouraged to *experiment* during the course of their work, and to savour sound for the sake of it; they must enjoy learning to 'taste' tone with their ears, like a discriminating winetaster who, rather than being bent on discovering his favourite red, learns to distinguish different qualities and characteristics in wines which are equally enticing.

If we do not dare to take steps – what are we afraid of? Teachers can tell when a pupil is taking even the slightest imaginative initiative in his exploration – the pace of progress quickens and the whole process takes on a greater excitement!

Imagery and hearing

Imagery, being a representation of something, is not the same as imagination, and doesn't have the same direct creative force. Since people perceive things subjectively from their own unique perspective we must understand that insisting on a particular image to describe a sound might be confusing or counterproductive. An image can only be as useful as it is clear in the mind of the imaginer.

Even generally accepted 'singing teaching parlance' such as describing a sung sound as 'forward', 'in the mask' or 'supported' can be confusing or misleading, since such descriptions are at best relative. All such terms carry connotations that may vary from mind to mind. In any case, if the sound you're making is described as 'forward' and therefore ' good', there has to be a way other than the physical sensation (which in this case if done well is negligible) for recalling it. If not, neither 'forward' nor 'good' will be productive.

It is far more productive and stimulating to aural perception to ask the singer to describe the sound he has made in his *own* terms. This can give him something of his own imagining with which to associate his sound and thereby with which to reproduce it. It's important to use descriptive terms – colour, character or emotion, for example – rather than critical terms. Knowing how the singer perceives his own sound can help a teacher to suggest appropriate moment-to-moment adjustments in the process as it continues. Describing his own sound can make it more 'real' and present to the singer. Remember, the fully enlivened voice does not stand still.

It is important to note that sound and feeling in the singer are frequently confused. For example, a singer will often say of a good but unfamiliar sound that it 'feels wrong': the sound is missed because for the singer it *feels* unfamiliar. I often have to ask again 'What does that *sound* like; are you really describing the *sound*?' Granted that both sensation and sound are subjective, the sound, issuing from the one source (the singer's throat) is objectified through virtue of the fact that it is the same sound sung and heard. This common aural ground must be sought, read and experienced if steady progress is to be made.

Sometimes we are listening but don't *hear* because our minds are fixed on some 'sound agenda' or model. We judge what we hear to be wrong because it isn't what we expected. This is an opportunity lost when so much in this process happens as if by chance.

Our ears in relation to sound must be like the eyes of an artist in relation to an object. If he wants to draw a particular house, it's no good drawing an imaginary one, or what he thinks it *should* look like. Terminology is also subjective and can be misleading. For example, people often confuse 'forward sound' with 'singing in the mask'; it's also described as 'open' (strange, considering what's actually happening at the cords!) The word 'forward', like all such labels, is in fact irrelevant. What must be agreed upon between the pupil and the teacher is the 'sound' that reflects the specific event.

Hearing versus physical sensation

The vocal events described above depend for their efficacy on other supporting events elsewhere in the body. This is why they're not felt where they are happening – they have been integrated. Seeking physical sensation somewhere else, however, would be misleading. So long as physical sensations are sought instead of the sound, a high degree of precise coordination cannot be achieved.

Naturally in hard training muscles are bound to be felt muscularly in, for example, the suspensory mechanism, and in various parts of the torso, depending on the focus of attention. The sensation is similar to any resulting from properly conducted muscular work, where there is deliberate relative isolation of the muscles involved. An important exception is the intrinsic musculature of the larynx, which depends for its success on adequate support from the suspensory system. Although these muscles must also be worked hard, they are 'protected': the suspensory muscles take the strain that vocal fold tension might otherwise put on the larynx. They set up the conditions in which work can be achieved without vocal harm or discomfort. This tells us that the extrinsic muscles must be well innervated if we're to have success with the complex mechanism within the larynx. A singer's throat should never feel tight, burning, tickly or hoarse as a result of training.

As muscles become more flexible and generally better integrated, a de-sensitising process takes place so that over time neither specific nor supporting events are felt any longer (except where appropriate in continued training). If the cords are stretching, closing and tensing efficiently, the body will behave accordingly. A finely-tuned pair of vocal folds signifies a finely-tuned body, and the effort involved is spread appropriately and evenly. The vocal system, which 'feels physically', perceives this integration as a whole, as does the ear. The only pertinent label left is 'singing'.

The description of any vocalisation (and the number of possible sounds and descriptions of sounds is infinite) continues to be valid as a teaching tool only so long as it has the effect of bringing back into focus the actual sound to which it was originally given. If progress is good, the description may need constant modification.

When we talk about and work with 'chest voice' and 'head voice' the labels signify *sounds*, which have their own peculiar characteristics and are inherently variable. The variability reflects the voice's human nature and its job as an expressive instrument. The sound of a singer's 'chest voice' per se will modify as it develops, and as it assumes its role, merging with 'head voice', 'forward placement' and everything else to form the balanced, flexible entity already outlined. Hearing the progress of this mixing and merging can be a subtle or slippery job, especially in the more advanced stages, where adjustments can be elusive because the difference in physical sensation is so slight. It helps to ask the singer when adjusting if she hears the difference in sound and what she did to achieve it. The answer often is: she heard but did nothing to achieve it! Even at this stage I might use the placing idea, suggesting 'a little more head voice or stronger *falsetto*'. The singer knows by now that I'm appealing to her aural imagination. Finally the ear takes over and physical references can be dropped.

Child's play

The state of play

One of the most successful and enjoyable ways of exploring and 'bringing out' a voice is through playfulness. For some adults, play can seem to be a regression, with the implication that beyond a certain age there is something wrong with it, and there's a risk of looking silly. Do children look silly? While we may see play as not serious, to be put away as we grow up, children have no problem with being playful. We never 'grow up' from the point of view of arriving at some predetermined fixed point. Robert Holden, author of *Laughter, The Best Medicine*, describes being silly as 'the first step to being free',[12.4] and states that silliness is needed for healthy, creative mental and emotional growth. Therefore it seems absurd to think that we can dispense with qualities commonly associated with 'the child' – curiosity, spontaneity, candour, exuberance and others – not the preserve of childhood but symptoms of life being openly lived at any age.

A 'legitimate' short cut

In learning to sing it is essential to break out of habitual or outmoded procedures or ways of thinking. Being willing and able to play with 'funny' sounds can be an effective 'short cut' to our singing voice, bypassing more formal, less free procedures. 'Silliness can suggest a new angle, a fresh perspective, lateral thinking and novel ideas', says Holden.[12.5]

The spontaneous movement in sound (the tonal quality and its continuity) is better *heard* by the singer's own ear if it bypasses the rational mind. This is highly instructive because it demonstrates how 'voicing' and 'music-making' depend upon different aural departments.

If a singer thinks 'I'm being asked to make this weird sound which doesn't seem to have anything to do with singing' he might well have misgivings. However, if he sets about it in a playful spirit without making value judgements, and with no reference to what he perceives as 'singing', he has a good chance of discovering aspects of his voice which otherwise might remain hidden because of habit, taste or how seriously he takes himself.

For example, simulating the sound of a large organ pipe or foghorn by making a windy 'vv…' sound may seem an odd thing to do in a singing lesson, but it might be more successful in beginning to open up a singer's chest voice than struggling to 'sing' a low note. Making the 'owl sound', 'te-wit-te-woo', (or 'te-woo') might produce just that elusive shift in alignment needed to achieve the perfect head tone which was not coming about through a mere 'oo…'! The possibilities in funny sounds as purposeful tools are endless. The worst result will be laughter.

Focus and discovery

Children's play is often highly focused. Children have the ability to concentrate and bestow value on the simplest things, and find endless fascination in the new. This doesn't mean that children are simple, but indicates a degree of absorption which tends to get lost as life becomes more rushed or complicated. Most voice-accessing devices are extremely simple and have a 'here-and-now' quality which thought can easily disperse. They also demand concentration and persistence. We can gain much from this 'simply focused', non-judgemental state, because with it we can more easily notice and accept discoveries which we might otherwise easily overlook or else dismiss.

Vocalising first

Getting over preconceived or prejuiced notions of 'voice' are crucial pre-requisites in our search for the voice which is genuinely our own, 'unconditioned'. Many aspects of the singing voice are better accessed initially without reference to specific musical pitches. When a sound, attained without a musical reference, does not then transfer at once to something more specifically musical, we learn about important differences between vocalising (the spontaneous), and music (the learned). If the freedom that a voice enjoys in the absence of specific musical references does not carry easily into notated music, this is because the mind interferes with the untamed spontaneous way of vocalising, seeking 'safety' with the 'known' controlled way. While musically this may result in accuracy, vocally there is no longer such freedom of movement (and therefore of expression or inventiveness) because the same rational monitoring system is being applied to the vocalising (and by extension to the expression of feelings) as to the musical performance. An academic approach to the development of vocalisation can more easily give negative feedback to a singer who, either in the teacher's eyes or his own, fails to 'get it right' than one in which it's not a question of 'getting it right' but of exploring possibilities. Incidentally, some of my pupils who have ventured into the demanding world of contemporary music have been thankful for a 'playful' upbringing!

The spirit (or genius) of the child is surely the spirit of humanity. In losing touch with its inspiring, self-realising and creative elements, we also lose the ability to communicate – to live in spirit – with others.

Laughter

Smiling and laughing are universal human expressions with which we convey a positive response to life. Laughter has many healthful benefits on all planes of our being. It stimulates the immune system and the heart, can relieve stress and pain, and is manifestly good for the spirit and more than useful in promoting tolerance and goodwill. Despite all these qualities, I once heard a statistic claiming that whilst the average daily laughter count for children is 400, for adults it is only fifteen!

Laughter can be seen to have suffered the same kind of repression as singing. There are certainly significant physiological parallels between the two deeply expressive activities. In learning to sing, usually after a lot of hard physical work and emotional agonising, the lines of communication between what is emotionally felt and what is heard to be felt become re-established. Accordingly the message gains clarity and force, and in a very physical way the emotions underpin and endorse the meaning of the sung words.

A side-splitting belly laugh triggers a powerful physical coordination between the diaphragm – 'the seat of emotions' – and the throat. Laughing (with or without specific pitches) can be an invaluable tool in training, re-establishing the strong and deep natural connections between the body and the throat and simultaneously releasing the diaphragm (which is often insufficiently responsive) and its relationship with the soft palate. We can no more usefully manipulate the diaphragm than we can laugh by numbers – it needs something to respond to.

Experiment 12.1

Take a deep breath and, holding it, try to laugh. Then try again, taking no breath, and see how easy this is.

When unimpeded, the diaphragm positively bubbles with excitement, and in due course fulfils its role as inhaler. Many people find a full-bodied, full-throated laugh difficult if not impossible. Whatever has inhibited this amazing device, we see that it works most freely and fully when in partnership with the suspensory mechanism. The larynx is thereby freed to make its strong 'stops' without strain. At each stop or stroke of the cords there is an increasingly strong counteracting movement (throat-opening tendency) which prevents explosive force. The stops are similar to glottal stops, but more fully 'cordal', nearer in fact to the *colpo di petto*. What is significant is the strength and flexibility between the throat and the body achievable on an out-breath with no prior inhalation and no force or damage to the throat. Laughing, like singing, is a spontaneous emotional response which allows no time for over-filling the lungs.

To sum up, laughing and singing share the following characteristics.

1. Strong support of the larynx by the suspensory muscles.
2. Strong reciprocal connections between them and the torso.
3. A free movement of the diaphragm mirrored by the soft palate.
4. Strong glottal closure without force.
5. At one stroke, or more accurately a series of strokes, mind, body and heart unite in an honest, vigorous and appealing (not to say contagious) way.
6. The lungs are replenished naturally upon relaxing.

Significantly, as the breath runs out, these dynamics 'tighten'.

Gestures, movements and inner rhythm

Few practices are more damning to singers and their performance than the insistence that they stand still while singing. Of course, there is nothing as distracting as a singer waving her arms about aimlessly, 'conducting' herself or swaying purposelessly. Nevertheless, movement is a natural response to the rhythm and lyricism of music. If a singer's movement is at odds with her singing we must find out why.

Standing still is in fact unnatural unless employed in a 'fight or flight' situation, when it will last seconds rather than minutes. To stand still, we must impose our will upon a structure wanting to renew its position. Naturally, the better the alignment of the body – and our relation with gravity – the less hard the hundreds of little muscle movements employed in keeping us upright and steady have to work. Trying to stand still encourages held breath, or at least a partially held system rather than a fully free and alive one. Standing still while being fully free to perform vocally or physically is a contradiction in terms. While moments of stillness in performance are sometimes indicated and desirable, they are usually short-lived, and, as with sustaining a quietly sung tone on a high pitch, need a highly trained body to be successful.

In a concert situation, standing still as a matter of course while singing looks as unnatural as moving for its own sake without relevance to the performance. In *The Thinking Body* Mabel Todd, who, like Alexander, was working at the beginning of the 20th century, states: 'muscles can be held in one position or continuously contracted only for short periods without fatigue... Motion allows for rhythmic alternation and variation in the use of muscle fibres, and gives time for the individual fibres to rest......'.[12.6]

Many singers are out of touch with their bodies, and therefore with their inner rhythm or pulse. Inner rhythm and the body's freedom to move well are interdependent. Singers' preoccupation with their voices often prevents total rhythmical involvement. This is particularly unfortunate because if the voice is not sufficiently flexible to respond to rhythm, a vicious circle is created which it is difficult to break. We cannot be rhythmical by 'thinking rhythm' – we can only join the dance of life that is somehow already in motion. Nowhere is it more important for a physical performer to find his inner rhythm than in singing. Since the singer is his instrument, the slightest bodily stiffness causes a chain reaction of structural constraint, which directly or indirectly affects all aspects of his performance. Being both the player of his instrument and the instrument of his playing, the singer has to *be* the rhythm, *be* the *legato* and the *crescendo* – he doesn't have to *do* these things technically if his voice is sufficiently liberated. It is when his voice isn't free that he tries to do rationally what should be intuitive.

One of the best ways of embodying the rhythm of a piece of music is dancing to it. This makes particularly good sense since rhythm seems to originate at the level of the pelvis, where the deepest muscles of breathing are situated. Moving the pelvis can be sensual and suggestive, and therefore for many people, consciously or not, out of bounds. Stiffness in this region seriously affects both breathing and emotional expression.

Singers who feel 'rooted to the spot' (as distinct from truly grounded) should be encouraged to move from the start of training. Getting familiar with, strengthening and enjoying movements below the waist can go a long way to freeing the whole extent of the breathing system (from pelvic-floor up), which otherwise tends to be put under stress from overwork. All movements must be carefully monitored by the teacher to ensure they are not contradictory in postural terms, that they are helping the singer to move internally, and are not merely externalisations or mannerisms.

The embodiment of a character demands the same degree of physical-emotional availability. Appropriate expressive gestures are more inclined to arise naturally from such freedom. Don't forget, characters have legs and feet as well as arms and heads! Emotions can also be released and embodied by dancing them.

Up or down?

Singers' gestures are often linked, not necessarily helpfully, to their vocal movements. An arm or whole body can make upward movements for musical intervals perceived to 'go up', and downward movements (usually bending knees or lowering the head or jaw) for those perceived to 'go downwards'. This tendency decreases as the suspensory mechanism comes into its own, balancing and supporting the larynx with sufficient strength and elasticity at all pitches. Meanwhile, much can be gained by deliberately making an illogical gesture (voice one way, finger, fist or body the other), providing synchronisation between gesture and voice can be achieved with ease and purpose.

Pat Moffitt Cook, in her preface to *The Rhythm Inside – Connecting Mind, Body and Spirit Through Music* by Julia Schnebly-Black and Stephen Moore, an excellent book about eurythmics, which is a highly stimulating way of improving musicianship, writes that: 'Sensory integration is best supported and expressed by linking auditory stimulation and body movement.' [12.7] The Greek word 'eurhythmy' means 'good flow' and can be applied to concrete forms such as sculptures as well as music. Jaques-Dalcroze, the inventer of eurhythmics,[12.8] talked of 'finding the flow', not only of music in the abstract, but that which united the musician bodily and spiritually with his music. 'Good flow' is achieved primarily through hearing. Moving to music is a response to hearing it. We embody music not through any intellectual process but by allowing our body to respond, to vibrate or pulsate in sympathy. Reading music is only the beginning of preparation for performance. We are only truly ready to perform when the music is 'known' by the body.

In training and singing, we usually find that the physical impulses made to produce a desired vocal result (especially when we're dealing directly with the small-scale events inside the larynx) are out of all proportion to what's required for a number of reasons:

1. A genuine lack of natural physical response.
2. The damning perception of relative musical interval 'size'.
3. Expectation of difficulties, resulting in disproportionate effort.
4. Lack of aural focus.

Small hand movements, such as making pencil markings on paper, can have a remarkable effect in scaling down the impulse while the singer retains both flexibility and concentration of tone. At the same time gestures can play a defining and definitive role in the realignment of our ears with our voice.

Encouraging a pupil to move fluidly and rhythmically reduces her tendency to strain or force, and facilitates communication because of the empathetic movement it induces in the audience. A singer's body should respond to the meaning and drama of a piece as easily as a good actor's.

Emotion is the excitement of feelings, which naturally seek expression in sound. Likewise, gestures can summon up strong connections between how we feel our feelings physically and how they are reflected in sung tone. Once physicalised, emotions are more easily recognised by the physical voice.

Experiment 12.2

Clench and raise your fist as though you are going to bang the table......hold it! Did you get a response from anywhere else in your body? Usually the jaw tenses simultaneously or the teeth clench. Your eyes may have changed their focus. The gesture is potent with feeling, even though you were only going to strike the table (without reason). Try again, imagining a reason for wanting to hit out. Now, keeping your arm raised and fist clenched (holding on to the mood) relax your jaw and teeth (breathing all the time)....you can do it! Now, whether you feel like voicing your feeling or not, your vocal organ is poised to express the feeling freely in sound. The sound a singer emits will be informed both by the gestures she's making and what they do to arouse the movement of feeling.

Experiment 12.3

With a smile on your face, thrust both fists up above your head............difficult not to feel something like triumph or exuberance. Try reproducing the same feeling with shoulders slumped and arms hanging limply. Try smiling with slumped shoulders.....?!

I'm not suggesting that a singer should perform 'Rejoice greatly' waving her arms in the air, although, as an exercise in rehearsal, such a gesture might lift this great exhortation to the appropriate level of excitement. Links between gesture and emotion are not automatic by any means, but intelligently used in training they can have remarkable results and save much time. Arbitrary arm-waving and swaying, on the other hand, can prevent us from tuning into the body's natural response to feelings.

In studio work, we have to find ways of tapping into appropriate feelings, not only because we must express them in singing, but because of the physiological connections between our feelings and their voicing. Listening to voices, one is struck by how (broadly speaking) different parts of the vocal mechanism seem to correspond to different emotional states. Although such an approach could deteriorate into a reductive 'emotions by numbers', expressing different feelings in sound can indeed stimulate different parts of the vocal mechanism.

I am aware that descriptions of sounds are subjective, that sounds in themselves are not emotions, and that such feelings as love and joy can evoke many different tone colours. By the same token, I believe that the amazing singing instrument has the capacity to be disarmingly clear and direct, and to profoundly move people without a word. We do well to seek out and encourage this unspoken side in training and singing. It tells us how directly from its source the voice comes, or it cries out (huskily or harshly) informing us how diluted, distorted or blocked it is. As listeners, we should cultivate the capacity, which we all have, to hear and feel what the voice (as distinct from any other aspect of the performance) has to tell us.

Music

We must put our vocal instrument in good order before we can 'play' on it. What training the voice, our 'self-instrument', means in practice varies from singer to singer. In most cases, a gradual transition from spontaneous informal sounds to those more brain-directed ones and from pure impulse to something which can reflect an idea without the voice losing its flowing integrity needs to take place before the voice can be benefit technically or be effective in performance. Such a transition might take place over many months or years or in small ways for short durations in the course of one lesson. We must recognise that the voice is not designed like a keyboard. To begin with, musical intervals (demarcated curves in sound) and rhythms (which are not the exclusive province of music) provide reference points only. In evolutionary terms, music emanates from the singing voice, not the other way round. In order to overcome the apparent discrepancy between spontaneous vocal sound and what is demanded by written music, the singer's ear must be highly trained, vocally as well as musically.

Students using musical material, however simple, for technical reasons should not get hung up on intellectual or musical aspects, judging their success by them. For example, singing music too soon, a singer may endeavour to sing in tune, disregarding the means she's employing to do so. This is a pointless exercise. On the other hand, the freer the voice, the more musical it becomes by nature. Musical phrasing begins with the shortest, spontaneous flowing movement. Longer phrases must be allowed to grow exponentially. The frustration of trying to be a performer, artist or communicator with a weak or imprecise instrument can be immense and even lead to vocal damage. At an appropriate stage in a singer's development, when she can be sufficiently spontaneous and not hesitate over notes and rhythms, simple formal music can begin to feed the singer's process.

Using music which appeals to the emotions of the singer and her lyrical nature can be far more productive than prescribed exercises. Simple phrases (from any songs or arias) with suitably ranging runs and musical shapes require minimal thought and no preparation. To begin with, tempi should be quick and rhythmical as slow tempi encourage hesitation and reduction of energy. Muscles need something to inspire them to dance! Simplicity and spontaneity are the names of this game; it must be as though the voice is inventing the shape, a lyrical context for itself.

This uncomplicated music-making can sometimes be introduced quite early on, but its aim is no more than to mark the beginnings of a transition from the informal to the formal. Rudimentary exercises may usefully be employed alongside the music, not by way of technique, (as in 'how to sing a phrase') but to stimulate muscles which are still coming along.

Working in this dual way – music interspersed with exercises – is likely to prove particularly necessary with a pupil who had a deeply ingrained unbalanced way of singing prior to training. A previous way of working associated with singing music will be remembered by the muscles that participated until the whole pattern is changed, the muscles re-programmed and the ear fully back in charge. This is why it is so important for this part of the process to be correctly proportioned and unhurried. Once the employment of music begins, devices that we call musical or vocal skills will already be familiar to the voice. *Legato* movements (basic flow), rapid little runs (vitalising the smallest muscle movements), *crescendi* and *diminuendi* (stretching and sustaining) and many other 'skills' will all have been released and developed as exercises.

Phrases carefully chosen from a wide range of music to complement the process already underway can be sung on suitable vowels to reinforce certain physical aspects of the voice and the idea of *legato*. Similarly, the sentiment or message of the text can be expressed to reinforce the lateral dimensions (expressive components) of *legato* singing. A strict ear must be kept on any adverse effects this 'intellectual' exercise may have on the integrity of the vocal structure, which must continue to be the basis of all future work. Only when this purely lyrical phase is well practised and familiar to the student, and continuity of line (preferably over her whole range) is assured, should words be attempted. At this point, it would be best to minimise the visual aspect by quickly memorising the words and rhythms of phrases before singing them.

The time eventually comes to start selecting whole songs, with the ongoing proviso that 'still difficult' bits can be left out or even temporarily rearranged. There is no point undermining confidence gained, at a stage where we are still exercising and gaining assurance and stamina. However, a certain amount of carefully chosen, challenging material is in order.

The timespan of this work (from the beginning of training to singing a complete song) can vary greatly from singer to singer. One of little talent is likely to spend many years in this progression and is unlikely to 'stay the course'. The so-called 'natural singer' may sing songs and arias from the outset but would be well advised to investigate and deal with even the smallest weaknesses, spend time fine-tuning his instrument and building up stamina, all the while educating his ear so that it knows what he's doing. I have seen many splendid voices founder because of rushing into as yet unsuitable repertoire. In between these extremes of vocal condition there's much talent in various stages of development. Training, if it is to be thorough, cannot be rushed.

Be bold!

Taking time with training does not mean starting slowly, with small aims in anything except duration of sessions. Boldness and exaggeration are of vital importance for anyone anxious to progress. Being tentative or hesitant will produce tension, not banish it. For the pupil, boldness means being prepared to jump in blindly if necessary, trying a new approach rather than sticking to familiar ways. The teacher should ensure that nothing harmful or irrelevant is attempted.

Teachers themselves may need to be bold, to be able to concentrate on certain vocal events (at the partial exclusion of others), and to learn how far to take a pupil with an exercise at any one time. If the training process is to move forward, the teacher must take calculated risks. As a teacher becomes familiar with a voice and its owner, the risks should become easier to calculate, and she will be able to gauge the strength of the work. Although nothing can be imposed on a voice, a pupil's exercising will always begin with something which the teacher hears to be available, even if to the pupil it is as yet unfamiliar. The teacher must learn how far she can usefully lead a pupil out of his 'comfort zone'. Usually this can be judged by ear, but in the hardest work one eye should be kept on the pupil, not only to monitor the effect the work is having on posture, but also to detect any visible signs of genuine distress. No physiologically sound vocal work should cause a singer pain in his or her throat. Following a strong contraction, however, muscular work may be felt to have taken place in specific extrinsic muscles and in the torso. Relaxation should follow.

Improvisation of means

Teachers need to be continually inventive and creative. While necessarily returning to similar ground many times, the work is never routine. Ears can record the minute changes which are usually the ones that count. It is therefore vitally important that these changes are heard for what they are. They can be further developed by similar or quite new means. When dealing with perceived improvements, we need to check each one thoroughly in order not to miss almost imperceptible backsliding. Singing music too soon can easily throw the physical development process as such into disarray without us realising it until we have regressed a few significant steps.

Ringing the changes

Ringing the changes helps to keep the training process alive and being creative can make it truly dynamic. The spiral symbolises the dynamic aspect of things. Changes of treatment are also indicated because of the influence one part of the voice may have on another. If the process is to make any sense, it is this influence we rely on. Working hard on a particular vocal process may be heard to be necessary. However, in structural terms we stretch the cords (for example) in order to accommodate tension, so we must constantly check that this is what is happening, and to what degree. We cannot make a plan of action with something as unknown and unpredictable as a person's voice. We need to allow for trial and error, with the important proviso that we constantly check and cross-check results.

Opposite contrasts

Considering the antagonistic nature of the voice, it is not surprising if occasionally we find ourselves needing to do the opposite of what seems required, realising that what a specific vocal event actually requires is its opposite number as training partner. Isolating vocal events for the sake of it is illogical and divisive. We work on one aspect so that it can blend suitably with others. Even if, early in the process, it seems that things have to be undone or backtracked, the various vocal events or strands should be drawing together.

Did I hear that correctly?

In singing, all the muscles involved in producing a sound stay involved in order to sustain it for as long as needed. In training, a specific sound event representing only part of the voice is required to sustain itself for a short duration. By the time the voice comes to phrases of music this particular event must be able to hold its own with all other muscular events, intact for the duration of each and every phrase.

In order to give his ears the best possible chance of monitoring his voice as it sings the singer must not look back, but keep going. In an exercise series, the teacher will indicate whether a sound is 'right' merely by encouraging the pupil to continue without hesitation so long as she considers it's profitable. If a sound was heard correctly, it can be reproduced without ceremony, because the ear will remember it without any mental analysis. While the teacher's job is to train her pupil's muscles, the pupil's is to be aware of the experience. With thorough training, muscle and aural memory will eventually be happily re-united.

In order to maximise work and minimise hesitation, when employing a stepwise series of exercises a teacher should aim for a 'good grouping' – a term taken from target practice, where what is considered good isn't so much where the target is hit as how closely grouped a series of hits are. A vocal exercise series doesn't have to be perfect to be useful. What must not happen is pausing to consider each gesture. 'Did I hear that correctly?' must be replaced simply by 'am I getting on with this?' The pupil must suspend judgement so as not to interrupt the spontaneous flow, the thing on which success most depends in this work.

In strong exercising, following and concentrating on the sound instead of the sensation ensures that you don't give in too easily or too soon to muscles' 'complaints'. The body may look for easier, compensatory options as it strives to keep going. The progress of the work is monitored by the sound, not by physical sensation, and sometimes to keep going the muscles of the torso have what I call a 'healthy struggle'. However, one should not keep going at any cost. An exercise is only healthy and valuable for as long as the sound says it is. The teacher must be listening to and judging the sound acutely. She must always bear in mind, however, that it is in the nature of the singing sound to change in flight to some extent. Keeping going regardless of a slight change may or may not be profitable. Radical changes show that the singer has strayed from the point of the exercise, and are often accompanied by a greater intake of breath and obvious compensatory muscle work somewhere in the system. The moment this begins, the singer's ear takes a back seat in favour of the physical sensation and the process is lost. A teacher relying on her intuitive ear must make quick judgements about the continuing usefulness or otherwise of an exercise.

Attention and distraction

Schnebly-Black and Moore (1997) write 'Attention serves as the bridge between perception and memory.' 'Stimuli may travel through the sensory system but not be perceived. Events may be perceived but not remembered.' They go on to say 'The primary factor in determining the level of retrievability is the level of attention operating during perception.'[12.9] Attention spans are often short. Experience shows, however, that they can be trained to last longer. It is also clear that muscles in the brain need rest as much as those in the body. The act of singing demonstrates that we can concentrate on several different but complementary activities at once. Our vocal training work is, however, distinct in the high demands that it puts on our aural sense. Add to this the muscular nature of the work and the extent of the muscle systems involved and we can see that the most productive training sessions are unlikely to be long. For a well-focused and motivated pupil who has a good idea of how to apply himself, 20 minutes is normally ample for this work.

Schnebly-Black and Moore write 'The emotional and motivational parts of the brain *including hearing* (my addition) known as the limbic system existed well before the development of the neo-cortex, the thinking part of the brain. Part of the sub-cortical structures, the limbic system was imperative for survival at a time when the world demanded immediate action without the luxury of time for decision-making.'[12.10] Concentration means being aurally *receptive* (not 'thinking'), something which cannot be achieved with a tense, struggling body. A general attitude of alertness is indicated, both for a spontaneous muscular response, and for general sensory reception. Once this is achieved, there are various ways of holding attention which include sudden changes of tack, novelty (playing) and contrast. Most of the modes of work described so far lend themselves to such stratagems.

A pupil may become over-involved with what's going on, perhaps due to the determination to 'get it right'. At this point, everything threatens to become tense and earnest. This may be the moment to talk about the weather or something else totally irrelevant, as a momentary distraction.

Confusion

Pupils 'get wise' in the sense of anticipating the work, 'knowing how this one goes' and enthusiastically demonstrating the fact, or anticipating when the teacher will end an exercise and therefore hesitating or losing momentum. Lazy bodies and controlling minds can conspire to find an 'easier' way. If things reach this point, it may be time to remind the pupil of her positive role in the work, or alternatively confuse her, so that she has to start finding her own way again. This may be a question of trying something more radical or of doing an illogical sequence of short exercises of no consequence.

Normally, of course, we are out to clarify rather than confuse. Often though, from the teacher's point of view at least, the work is going well and a pupil will declare 'I don't know what I'm doing!' This should be taken as a good sign – she doesn't know *physically* or cannot explain it *mentally*, so has let go of inappropriate goals or controls. When this happens, it's important to encourage the pupil, finding out how she feels. She could be disconcerted or even embarrassed at 'losing control'. She may think that she's done something wrong and wants to 'grab her props' again. While a pupil has to learn to live without props of control, feeling a loss of control has an emotional component, which can be painful and must be handled with care.

Sequences and the order of events

The voice is a complex instrument which, in singing, has to contend with words, musical lines and emotions while demonstrating numerous vocal skills. Logically, the better working order the voice is in, the better chance one has of singing well and the less need there will be for artifice or for stratagems to get round difficulties. Teachers often have to attend to all the facets of an aspiring singer's art and feel under pressure to constantly produce a tangible or measurable result. There is, however, only so much that can be done in the given time and it is always more beneficial to follow a logical sequence of work. Suppose you have one hour. Spend perhaps ten minutes on relaxing, energising and warming up, paving the way for singing as such. Then spend, in a purposeful, structured fashion, the next half hour or so on text and music, remembering that quality is always better than quantity. Finally, spend up to fifteen minutes (no less than ten) on hard physical, vocal training. The pupil can then rest his voice and his muscles should benefit from relaxation.

Exercises are more likely to be done sequentially than in isolation. Beginning in a certain direction and capitalising on the results at each turn, the various purposeful physical manoeuvres sometimes reveal what you hoped they would. At others they reveal something unexpected which can often be used productively. While there are tasks to complete, we may not be able to go straight to them, and may have to try different routes at different times. An obvious problem may lie deeper than expected. The *near* cause, the one that's easily seen and criticised, may turn out to be merely a symptom. Symptoms are warnings, telling us that something is wrong that we cannot see, and merely relieving them is deceitful. Such work will eventually be shown for what it is, when different symptoms arise from the same deep source.

Starting points

I once heard life described as a river, not a stone, implying that there are no predetermined beginnings and endings. There is wisdom in this from which we might take comfort, for nothing in life seems ever to be quite finished or complete. Would-be singers arrive at the studio or college in very different conditions and their training should by rights begin from their individual positions and conditions and be open-ended.

Although training work should be structured and progressive, there are no hard-and-fast rules about where to begin a session. The voice leads the way. The moment a singer opens his mouth we have an indication of how we should proceed. If in doubt, we can adopt a checklist while doing the 'vocal rounds', noting what is strong or dominant, and what is weak or in need of investigation. Neutralising the throat and humming, which leads the voice away from faulty speech patterns, are usually indicated at the start of a session. Two obvious logical muscle training exercise sequences are:

1. Exhalation, suspensory mechanism (capitalising on the in-breathing reflex), larynx (depending on adequate support from the suspensory mechanism).

2. On the principle of working from strong to weak, or big to small: chest register (on low pitches), edge mechanism (much finer version of chest register, most easily accessed in the middle range), closers (capitalising on the refined strength of the edge mechanism).

Types and shapes of vocal sessions

The purpose of the voice-work being undertaken should always be clear. Distinct from a pupil's own preparation, which will be discussed later, there are three basic types of work.

1. Hard training

This means working on the muscular structure of the voice, with the purpose of re-integrating its many parts until it resumes its organic shape and self-dependency. This systematic work covers all areas, ranging from the large muscles of the breathing apparatus to the smallest fibres of the edge mechanism. The work must be well-apportioned and gradually more meticulous. Suitable periods of relaxation must be carefully judged. As a general rule, a good session of hard training consists of a series of *crescendi* of exercises (in terms of strength and duration) followed by a *diminuendo* by way of 'cooling down' at the end. Following the session, the singer should rest his voice for a period commensurate with the length and strength of the training session.

2. Working on music and text
This should be well planned, with a strict ear kept on vocal productivity. Many singers think that the more they sing, the better they will do it. This is a dangerous illusion. Up to a certain stage of integration, *how* you sing is more important in terms of development than *that* you sing. Suitable breaks for muscles to rest should be made, and these can be used to examine text, listen to harmonic sequences, study awkward rhythmic passages and so on. Working with music and text should always be purposeful and vocally constructive.

3. Warming up
This is not a workout, and its purpose should be to arrive at a point, mentally as well as physically, at which singing follows on naturally. A singer may or may not need to warm up before singing. When a voice is in good shape, singing is as easy as walking. Often suitable music can be as good as if not better than 'warm-up exercises'. A vocal warm-up is the equivalent of a dancer limbering up so that his muscles aren't 'cold' when the dance begins. It will include finding the right energy level and frame of mind. Any vocal exercises should be based on the condition in which the voice finds itself and not mindless routine. If a singer is not sure how to warm up, it is best to sing something musically undemanding, enjoyable, lyrical and alive. In preparing to sing, it's always helpful to remind oneself that singing is an activity that depends on something to respond to and to inspire it. A technical warm-up doesn't necessarily prepare you in this way. Recalling the mood of a piece of music, going through the words (not so much for the sake of memory as to attune yourself to meaning, sub-text and sentiments), and getting into character can lead to a state of mind the logical outcome of which is to sing.

Caricature
I sometimes encourage a pupil who is reticent about making a strange sound to caricature my demonstration. This often does the trick because the pupil doesn't feel entirely responsible for the outcome. It is important to be sure, however, that it's the sound that is captured, not the physical sensation or funny face!

Tensing the tense
'Legitimate' vocal muscles, when over-working, can be brought into line by training their opposite number (antagonists). A knotty problem caused by compensatory

or irrelevant habitual muscle contractions can sometimes be resolved by tensing the tense muscles further and then letting them go. This makes way for 'legitimate' muscles to have their say. Since this device is employed only for a second or two, any feeling of discomfort will be equally short-lived. However, remember this is not an exercise and should only be employed in extreme persistent cases.

Pupil mimicking

While this too should be a last resort, sensitively done it can save time when you know a pupil *can* make the sound you're after but for some reason does not, or thinks she's making it but isn't. It is simply a question of saying 'Let me show you what you're doing, and then you can compare it with what I'm asking you to do'. The comparison, once described, can be aurally instructive. The main problem here is that spontaneity is likely to be replaced by unnecessary pondering.

Singing in the dark

I have conducted tentative experiments working in darkness. This has had the effect of concentrating pupils' aural attention. Trying exercises with closed eyes can also have this beneficial effect, especially in those who are inclined to scrutinise the teacher in order to ascertain how they are making a sound!

***Attacca*!**

This is a misleading term as it carries brutal connotations which should not be linked to singing. Even the word 'onset' is described in my dictionary as '1. attack and 2. A beginning, esp an energetic or determined one'.[12.11] While the inception of a sung tone should be energetic and determined, it should by no means be brutal, crude, forced or explosive. Beginning a sound cleanly is a problem which will only be resolved when the various closing mechanisms of the glottis and their antagonists are highly trained. Two types of onset are the so-called 'hard attack' (or glottal stop) used for example at the beginning of certain German words, and the so-called 'soft attack' which is aspirated. The ideal start to the tone is a clean, decisive attack which is neither hard nor soft. This should be cultivated from the start of training and used always except when either of the others is required.

A useful analogy might be the application of a bow to the strings on a violin or cello: loosely or too lightly applied, the sound will be 'fuzzy' (breathy), while too strongly applied it will sound 'scrunched' or 'explosive'. A string player must hear and in her case feel the exact combination of contact and movement if the onset is to sound neat and at one with the rest of the phrase. For a singer the equivalents are glottal adduction and impulse. In singing it is vitally important to be fully 'on' or 'in' the tone the instant you begin, in order to communicate meaning from the beginning. In other words, the successful onset is rhythmically spontaneous rather than technically prepared.

A skilful start to the tone embodies the virtues of both soft and hard attacks: the sense of inevitable movement that accompanies an aspirate, and the firmness or tonal point of the hard attack. Prior to a deliberate soft attack, the complete tone should be imagined so that the 'h' has no time to get in its way, or to dilute or 'press' the tone. After the instant of the hard attack and their tiny moment of prominence, the specific closers resume their vital contribution to the tone in due proportion to the rest of the ingredients. They are employed with emphasis but without losing integration with or disrupting the integration of the other vocal components. Only in high staccato singing (as in the arias of the Queen of the Night) do the closers go out on a limb and the voice turns pale or dry, manifesting as a result the sound termed *voce bianca.*

Coup de glotte

The exercise for activating the closers is called, literally, 'stroke of the glottis'. This stroke is no cause for concern unless executed with explosive force. On the contrary, being such a necessary and important part of the whole vocal mechanism, the closers must be exercised. While it's true that 'dry' consonants such as 'd' and 'b', (in conjunction with, for example, the vowel 'ee' – Italian 'i' –) produced rapidly and dryly by the tip of the tongue and lips respectively can activate and exercise the closers with no threat of glottal explosion, the use of consonants can also cause the larynx to rise and fix upwards if the openers are not sufficiently active. Much preliminary work might be needed to obviate false antagonism.

Aurally, and for the purpose of *locating* the closers accurately, the *coup de glotte* by itself can be more facilitating. I almost always begin work on this with the sound of the mechanism without vocal tone by way of locating it. This can help to dispel the notion that the *coup de glotte* is at all explosive. Not deliberately making a vocal sound, there's no struggle providing the throat is neutral. We are simply putting the cords together with the least amount of persuasion. The subsequent 'click' is the sound of a minute quantity of air eagerly filling the little vacuum between the cords on their release, like removing the tongue from the roof of the mouth. This 'stroke' gives us a good idea of how delicately the cords can close. In order to hold their own with everything else in the vocal mechanism, and be able to perform with the greatest agility, the 'closers' need to be strong in their own right and matched in strength by their opposers, the 'openers'. The 'closers', together with the edge mechanism, constitute what we call 'forward tone'. In singing, these mechanisms ensure that the singer's message is direct. 'Covering' techniques prevent this directness.

Inhalare la voce

Inhalare la voce is an instruction that helps us to sustain the relative stretching and opening of the vocal folds during singing, so that the closing and tensing events can take place with due freedom, strength and accuracy. The resulting antagonism is like the drawing of a bow – the employment of opposite elastic forces. It is this that facilitates the swift flight of the sung tone, something that would be altogether less effective if the arrow (tone) were simply thrown at the target (audience)! The analogy with a bow suggests both the strength of the archer and the fine balance of forces involved. This balance and strength is precisely what a singer's body needs if accuracy and precision are to be achieved.

In the course of exercising, the in-breathing quality of the tone must be exaggerated in order that those muscles which deal with the opening of the throat gain sufficient strength. Remember, open-throatedness is the result of stress being taken off the larynx by the suspensory muscles. The full stretching of the vocal folds (full head voice) is achieved by this indirect means. But for the inward-drawing that *inhalare* implies, the singing voice would sound shrill or hard, alarming rather than attracting.

It's important to remember that *inhalare la voce* refers to antagonistic muscular events and not to breathing in as such. When we breathe normally, the throat opens sufficiently for the purpose. In yawning too it does what it needs to do, if a bit awkwardly or reluctantly. However, normal, weak throats are not accustomed to opening freely to the extent required for singing. In the course of the sung tone, the breathing-in apparatus is continuously on 'high alert', so that at the end of a well-vocalised, vocally balanced phrase, when all that is required is to stop and let the body recover, breathing in is automatic and unimpeded. The body naturally returns to the state of having two lungs full of air. Manipulated or 'held' breathing does not allow for this natural recoil.

Note that in this process, the false cords are drawn apart (see illustration on page 40). This ensures that they will not try to interfere with the singing when the glottis is closed to produce the concentrated tone. In singing, this in-breathing quality does not preclude the necessary anticipation of energy required for following phrases.

Colpo di petto

Like the *coup de glotte*, this strong coordinating device can easily deteriorate into something explosive or pushy. In fact, it has little to do with pressure of breath, and less to do with pushing or manipulation. The *colpo di petto* is useful because it can have the effect of synchronising at one compact stroke practically all the muscle movements that constitute the sung tone. (I say practically, because, at least to begin with, it is less refined than the ideal onset of tone that our meticulous work is aiming for.) Vocal fold stretching and tensing are the strongest ingredients in the *colpo di petto*, and the inner chest muscle plays a strong supporting role. This strong training device, which is more like a vocalised stroke or striking behind the sternum than a stroke of the 'chest' (which suggests something too general), is executed without breath and should not be attempted until the extrinsic muscles of the larynx are sufficiently developed. Once a singer is strong enough in *this* area (not simply able to get away with misguided pressure against the throat) the *colpo di petto* can help both to consolidate the coordinated impulse of singing, and to gauge the strength of antagonistic forces throughout the entire vocal apparatus.

Appoggiare la voce

This translates literally as 'lean the voice'. As with 'supporting', 'leaning' invokes an image of one thing up against or depending on something else. The term '*appoggiare*' can therefore easily be misconstrued. In singing, there seems a fine line between a tone which is a little too 'giving' or 'loose' and one which is inflexible or too 'reined in'. *Appoggiare la voce* can help us experience the sung tone as something both flowing and firm, that vibrates freely but stays rooted. Again, the centre of physical attention is behind the top of the sternum (certainly not any higher, and not down where the inner chest muscle is located). This seeming 'centre of events' becomes quite specific and concentrated as training advances, and probably becomes the only really useful physical reference in singing. There should be no tightness, no resistance, just a comfortable, firm, secure feeling of being grounded, centred or 'at home'. This coordinating device is open to abuse because it shares similarities with the *colpo di petto*, and can come to be considered as a 'magic button' to set everything in motion; this is almost bound to cause tightness in the neck or force from below. In attempting to find this imaginary centre of events, we should understand that it *comes about* as the result of a comprehensive balancing process and is the manifestation of balance, not its cause. A well centered, well grounded voice can bias itself or lean out to extremes of the vocal structure with ease. A good example of this (with considerable risk-taking) is the singing of the great tenor Beniamino Gigli whose 'flirtations' with *falsetto* are so extraordinary that at times you cannot be sure whether his voice is still mixed or not.

Conclusion

In theory, there can be as many enabling devices and procedures as there are moments in each individual's vocal regeneration process. While specific outcomes might be desired and imagined, they are by no means automatic. The process is therefore one of trial and error, in which the teacher's ear continuously questions the singer's voice and evaluates the answers that it gives. Each answer is a new insight which invariably poses new questions. Thus the process proceeds.

PART II

Chapter 13

Muscle Training and Fitness

Our ever-increasing dependency on mechanical aids for comfort and efficiency has meant that for a long time we have made less and less physical effort in everyday life. Benefits of increased comfort and economy of effort are largely an illusion, since inactivity brings on ill-health. Loss of muscle tone makes for poor performance in the simplest of physical activities, and adversely affects the quality of our life, our sense of well-being and sheer aliveness. When it comes to something requiring unusual strength (such as climbing a steep hill or lifting a box of books) we often unwittingly force the issue because of diminished strength in one area or another. If, in singing, the various muscle zones are not well-coordinated and balanced in strength, strain is inevitably put on the breathing system, the spine, the thorax, the throat or the larynx itself, depending on the nature and severity of the imbalance. A reckless combination of imbalance and enthusiasm is evident in the way many singers approach their training or practice, with mindless and often forceful repetition of exercises that have little or no physiological basis or purpose. Negative effects, especially long term ones, are not always immediately discernible. Repeated muscular activity, correct or not, can lead through muscular desensitisation to a feeling of rightness and security.

To train voices successfully in a genuinely progressive fashion, teachers must learn how to work with muscles and how to monitor this work aurally. It may be comforting to know that all the muscles with which we are dealing have the same structure, and all function and develop in the same way, by contracting and relaxing. We induce vocal muscle work through sound. Beyond this, it is useful to know that the way these muscles are designed makes for:

1. Strength – the ability to *sustain* contraction
2. Flexibility – the ability to operate through their full range of movement
3. Coordination – the ability to co-operate smoothly with other muscles for a variety of purposes.

To these ends, muscles are subdivided into muscle bundles (the largest component) and muscle fibres (of which there may be few or hundreds, depending on the specific role of the muscle). Furthermore, each muscle fibre has a regularly ordered sub-cellular structure. Knowing this can help us to imagine the scale of the work we are concerned with at any given moment, whether it be relatively large (as in the extensive breathing apparatus) or tiny (as in the vocal folds, with their tapering edges). The overall aim with all muscles we are re-activating, developing or re-conditioning is similar:

1. Ease of operation
2. Optimum performing capacity
3. Optimum strength
4. Flexibility (elasticity)
5. Sustaining power.

Ultimately we are aiming at optimum overall efficiency.

The energy source of muscles is obtained by two basically different types of exercise, depending on the biochemical properties of different muscle fibres:

> **Aerobic**: prolonged, sustained exercise (such as occurs in walking, swimming or cycling) during which the body's absorption of oxygen is increased. Aerobic-type fibres are fatigue resistant and contain a relatively large blood supply.
>
> **Anaerobic**: short bursts of activity (as for example in weight-lifting, sprinting or archery) which require little or no oxygen. Anaerobic-type fibres tire easily and have a limited blood supply.

A third type of fibre contains the properties of both the above.

Muscles take care of themselves, in the sense that they know which fibres to use for which activity, and therefore whether or not they need more oxygen or less as soon as activity begins. This is particularly important when a body's muscular activity is constantly changing. Out of condition muscles must be broken in gradually.

If we experience difficulty in, for example, cycling, we say we are 'not used to it'. This means the specific muscles needed are not used to this activity. When out-of-condition muscles are worked for too long we get 'puffed': they are short of oxygen. A person's fitness can be measured by his or her ability to consume oxygen. Somebody unfit consumes half the amount of a fit person. The ability to consume more oxygen, and with it achieve greater performance efficiency and endurance, is gained over a period of time. The performance of a physically fit person requires less energy. While general fitness (the responsibility of the pupil) may not produce a good voice, it can make training specific to the voice (the teacher's responsibility) considerably more efficient and effective.

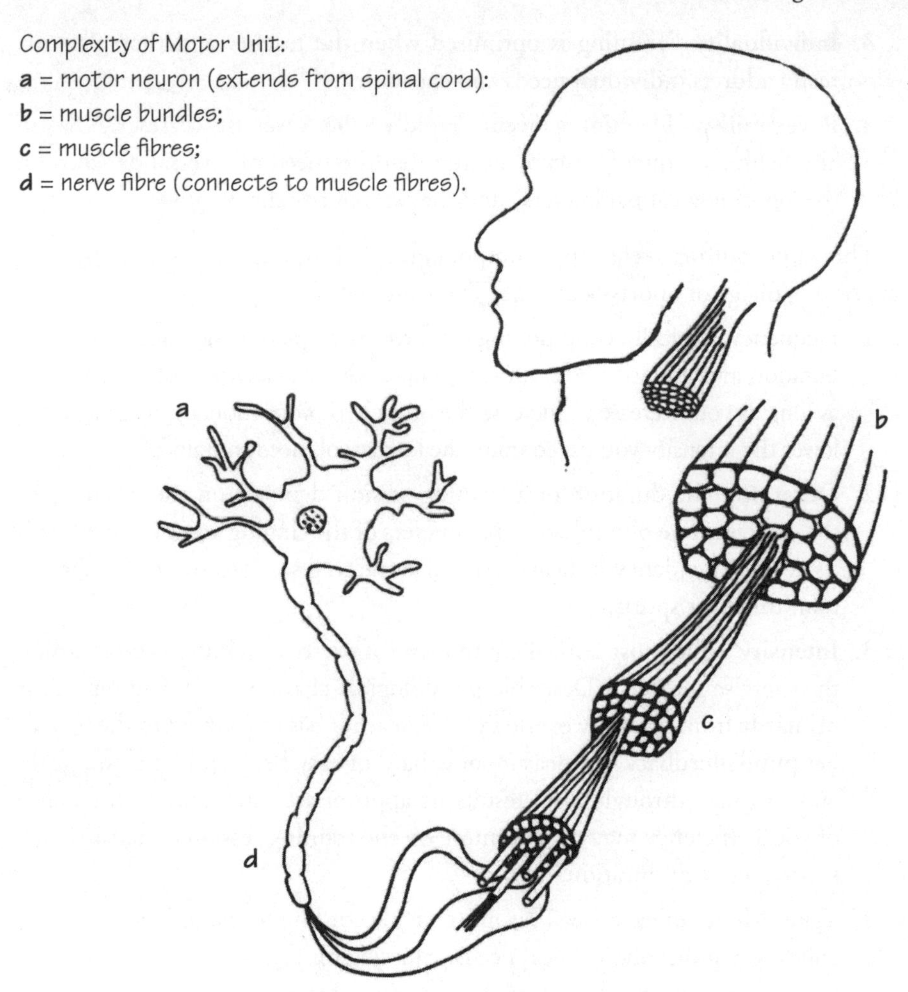

Taking care to exercise muscles appropriately, to quote Saxon and Schneider (1995) 'delays fatigue, increases duration and limits injury'.[13.1] These authors list four principles of training developed by exercise physiologists:

1. **Overload.** They write: 'Stated simply, the *overload principle* says that you cannot train a muscle without demanding more from it than it is used to giving.'[13.2] Needless to say, this demand must be very carefully gauged, discontinuous, and judiciously increased according to a gradual growth in natural capacity.

2. **Specificity.** In this case, this term signifies that we are working for singing not ballet or weight lifting. Saxon and Schneider write: 'The overloading aspects of a training program should coincide with the way the muscles ultimately are expected to perform.'[13.3]

3. **Individuality.** 'Training is optimised when the teacher and therapist correctly address individual needs and capacities'.[13.4]
4. **Reversibility.** 'Detraining occurs rapidly when exercise ceases'.[13.5] Saxon and Schneider qualify this by noting that the speed of reversal depends on the 'physiological parameters', and that loss of benefits varies.

The same authors refer to 'Components of Training' established by The American College of Sports Medicine. These are:

1. **Frequency.** 'Basically, conditioning is a three-part equation including frequency, duration and intensity. You can vary – up or down – any part of the equation as long as you decrease or increase the other two proportionally. In general the lower the intensity you use to train, the longer you need to train.'[13.6]
2. **Duration.** The duration of a training session depends on the intensity of the exercise. The physiological parameters of the singing voice coupled with its aural dependency indicate that voice training sessions should be shorter than those for sports.
3. **Intensity.** 'The most critical component of exercise training is establishing the exercise intensity. Desirable physiological changes from training occur primarily from intensity overload.'[13.7] A teacher has to learn, with the help of her pupil's feedback, to measure how hard or easy her pupil finds the work, and to gauge through aural results its appropriate intensity. 'Maintenance of vocal efficiency may depend more on the training session's intensity than its frequency or duration.'[13.8]
4. **Type (Mode) of exercise.** This must vary according to the individual's vocal condition and evolving vocal needs. Muscle groups can be worked on separately or in combination throughout the vocal system.

Not only must we distinguish between training the voice and actually singing with it, but we must be aware of the similarities, so that we're not just training in a vacuum. Singing resembles sprinting in the sense that sung phrases can be short, but also resembles swimming when it comes to singing phrase after phrase. When, after sufficient balancing work, a voice starts coming together, it can manage to sustain literally only two or three notes in a true *legato*. As it gathers sustaining power, these notes gradually extend to a whole phrase, and finally to a series of phrases. This is not unlike the progress of a high jumper raising her bar by centimetres.

Singers cannot go by the same indications as sports trainees, because there are too many personal variables and because of the crucial aural factor. Singers are inclined to exercise by feeling for what they cannot see because they have not learned how to hear it. Worse, they generally don't know how to reconnect to the natural aural monitoring device. As a result they bring an altogether too cerebral approach to their work. How can a singer make headway? The answer lies not in how she monitors what she's doing – aural perception has to be regained over time – but in how she approaches the work. The following are the best ways to approach the training work that I know.

1. Spontaneously

If there's a 'most important' way of approaching vocal exercises or any vocalising it is without thought of 'how to do it'. It is in the nature of the singing voice to burst into song without making technical preparations. The natural singing voice doesn't know about technique and doesn't want to know. Training this voice must take into account its impulsive nature and that primarily it responds to the need to express emotion.

2. Rhythmically

Husler and Rodd-Marling remind us that as in all forms of organic life, 'muscles are rhythmically constituted'.[13.9] In other words, a muscle without rhythm is a dead muscle. Sluggish muscles cannot coordinate with each other. A well-coordinated instrument is characterised by its ease of continous movement, because it is rhythmical throughout its structure. To bring a sluggish and poorly innervated muscle back to life we must appeal to its intrinsic rhythmic quality. We do this by exercising with a *rhythmical impulse*. In training, slight changes of rhythm can result in dramatic improvements in muscle response.

3. Rapidly

It is common in vocal training to think it logical to progress from something slow to something quick. This is either the result of mixed objectives, or ignorance about how muscles work. In music 'fast' *looks* more difficult. After all, the slower the pace of the music the more time there is to think about how to sing it. However, if we understand that it is more difficult for muscles to sustain their contractions than to make short ones, we have to reconsider how we treat them. Muscles don't need oxygen to set themselves in motion, but they do need it for sustained work. Muscle training begins therefore with short bursts of action (primary movements), which also serve to 'prime' the muscles with oxygen. Gradually the activity is prolonged. The same applies not just for linear endurance but for venturing 'higher' or 'lower' in pitch, both with partially isolated vocal events and with combinations. Rapid upward or downward movements will get a voice there sooner and more easily than slow deliberate ones, with which it may never arrive.

4. Strongly

It hardly needs saying that unless a muscle is contracted strongly enough it won't develop. Try making a fist without any strength; you're contracting the muscle movements that go into making a fist and yet barely notice it. Make it a little more strongly, and you know you are making a fist. Now make a fist as strongly as you can....and release it! Doesn't the blood simply pour back into those knuckles? After recovering (fully relaxing) doesn't your hand feel a little more energised? The first fist represents a contraction with minimal energy, the second one with purpose, and the third an unusually strong contraction. In training terms, the third does the most good. However, we have to be careful not to contract muscles *too* strongly, especially ones not used to making such an effort ('short bursts' does not mean violent!), or for too long at a time. We are interested in flexible strength, not brute force. Strengthening work must maintain a sense of release, for it's the release following their contraction that brings the benefits to muscles. Weak muscles are the ones that need training. There's no virtue in prolonging the effort of muscles which are already strong, or relatively so. In the long term, we are learning to use minimum effort in all our movements.

5. Energetically

Pupils sometimes labour unnecessarily, not for lack of capacity or will but for too little personal involvement, as though they are treating the body as something detached from themselves. This is not in the nature of the instrument, while lively coordinated movements are. Energy is better facilitated and more effective when pupils 'own' what they're doing as though the pleasure is in the possession. Entering into the work heart and soul improves its quality, because it is no longer effort for effort's sake.

6. Expressively

Once a pupil gets over the embarrassment of being expressive or overtly emotional with vocal sounds, bringing character and meaning to them, better physical connections are made between the 'department of emotions' and the voice itself. It is because these work so naturally and intimately together that shyness or vulnerability is felt. In singing, the more the voice is fed with emotion (its most nourishing food) the better it grows and the sooner it recognises itself for what it is. We begin to *know ourselves* in our sound. Even the shortest and simplest vocal gesture can be touchingly and powerfully expressive and thereby truly enlivened in singing terms. In training, this injection of meaning immediately raises the work to a new level of productivity, assisting both the direction and continuity of the impulse and equating physical events with tonal character.

Adopting these six ways of working, as a general rule, goes a long way towards avoiding the pitfalls of exercises that are repetitive and goal-oriented. It facilitates genuine long-term progress. The pupil's input is as crucial as the teacher's feedback if aural and muscle memories are to keep pace with one another.

Exercise progression

Serious athletes make it their business to be fit and they understand the necessity for regular training and sacrifice. They are motivated to improve their physical performance. I have rarely come across this single-mindedness in a singer, whose job is in so many ways more demanding. Much of what singers do or don't do in between training sessions has a significant bearing on their *ability* to make progress, even to the point of cancelling out the physical progress made in lessons. This is especially true early on when they're learning how to be appropriately disciplined in their health as well as their work. A physically unfit person makes much slower progress than a fit one. In terms of specific exercises this means that the rate of increase of exercise intensity may be too slow to make significant overall progress. The aural nature of the work means that little is to be gained by unsupervised exercising. By the time a singer is knowledgeable and disciplined enough to work on her own voice she can usually benefit more by simply singing.

PART II

CHAPTER 14

Singers' Health

Introduction

I doubt there is anyone more prone to occupational health hazards than the singer. The problem is that they are lurking everywhere: at home, in the open air, in performing spaces. If you want to discover just how widespread and varied the hazards of life are for those who depend on the health of their voices, consult Robert T. Sataloff's tome *Professional Voice,* which covers every conceivable vocal ill and its clinical care.[14.1] A big problem with health in general is that we spend too little on prevention of illness, preferring, it seems, to wait fatalistically until something goes wrong before seeking a cure. Although one might expect a budding singer to be fitter than the average person, because what he does is in itself good exercise, I would estimate that 80% of singers are less fit than they need to be in order to train and perform to their capacity. Exercise for both general and specific fitness should be for the long term.

Below, I outline some of the areas of basic health maintenance that I consider crucial for singers.

Ingestion

Fitness is not achieved by physical exercise alone. Materially at least we are what we eat, which means that every cell of our body from muscle to brain depends on getting suitable nourishment. Naturally the most nutritious food is uncontaminated and comes from 'clean' soil. Organic food production recognises that we are from and dependent on nature. We only tamper with nature because we don't trust it, because it has broken down through lack of care or because money can be made by exploiting it. Here we have an interesting parallel with the treatment of the singing voice!

In addition to ingesting chemicals through processed foods we absorb them into the blood stream from skin applications such as certain creams, cosmetics and shampoos. Our bodies do not have the wherewithal to deal with such a chemical siege; witness the rise in new disorders of the nervous and autoimmune systems. Even many medications have negative side-effects.

Digestion

It makes sense to discover what is best for our individual metabolisms (the natural chemical process that breaks down the food we eat so that energy can be created by our body cells) by consulting a nutritionist. It is his or her job to determine and deal naturally with individual intolerances and allergies. Apart from the vital energy factor, I have noticed that singers often show too little concern for digestion. A stomach full of undigested food seriously affects breathing. In addition this food is not yet available as energy.

Eating regimes, like metabolisms, are individual affairs. The great tenor Beniamino Gigli is reputed to have gone without food for six or more hours prior to performing. Generally, I would recommend eating at least three hours before a performance. Saxon and Schneider advise 'That pre-performance meal should consist mainly of carbohydrates. Besides being a primary source of energy . . . carbohydrate is also easy to digest and a good source for blood glucose concentration . . . fat and protein take 3 to 4 hours to digest. . . . You should be comfortable with the pre-performance meal.'[14.2] In general our diets should be well-balanced and contain plenty of fibre.

Weight

Sensible pronouncements about weight are concerned with fitness to do the job in hand. Saxon and Schneider state 'A balanced diet and maintenance of ideal body weight are significant variables in the achievement of physical fitness'.[14.3] The terms 'overweight' and 'underweight' (also an unhelpful condition for singers) refer to the fat-to-lean ratio of our body composition. In our highly physical context, fat extra to our body's needs is an unwelcome burden to the many specific muscle systems, which need to coordinate with balanced ease. In particular, the back must work overtime to carry dead weight and the delicately balanced strength and flexibility required at the level of the diaphragm is impeded. Singing being an athletic pursuit, the idea that extra fat helps is an unfortunate myth.

Medically, obesity is considered an illness. In *Professional Voice*, Robert Sataloff lists problems associated with it. These include psychological stress, high blood pressure and the risk of developing certain respiratory problems. Sataloff writes 'In the singer, weight should be lost slowly through modification of eating and lifestyle habits. Loss of 2 or 3 pounds per week is plenty. More rapid loss of weight causes fluid shifts that may result in changes in vocal quality and endurance.'[14.4] Serious loss of weight should be accompanied by purposeful vocal training.

Water

The human body is about 75% water. It is not unreasonable to deduce therefore that 75% of what goes wrong with our health is water-related. With 2 million sweat glands contibuting to all our other liquid secretions we can see how easy it is to become dehydrated. Saxon and Schneider advise 'drink beyond your thirst mechanism. This means to drink water after an activity even if you don't "feel" thirsty'.[14.5] Dr F. Batmanghelidj, in *Water and Salt*, writes 'Dry mouth is one of the very last indicators of dehydration of the body. By the time dry mouth becomes an indicator of water shortage, many delicate functions of the body have been shut down and prepared for deletion . . . a dehydrated body loses sophistication and versatility.'[14.6] He estimates that by the time we feel thirsty by this indicator our bodies are already dehydrated by two or three glasses of water, and it may take hours of regular water drinking before the body is fully restored to a healthy state. Topping up just before performing does not radically alter the situation if the singer is already dehydrated, and can cause discomfort. A dry mouth can be lubricated with chewing motions or by imagining biting into a juicy fruit! Incidentally, sometimes when we feel hungry our sensations are confused and we are actually thirsty.

Dr Batmanghelidj believed that it suits the pharmaceutical business to ignore or play down the vital importance of water. Among 46 reasons why your body needs water every day, he points out that water greatly increases the efficiency of the immune system, helps reduce stress, anxiety and depression, integrates mind and body functions and helps reverse addictive urges, including those caused by caffeine and alcohol.[14.7]

It is important to realise there is no substitute for water. Dr Batmanghelidj writes 'as far as the chemistry of the body is concerned water and fluids are two different things'.[14.8] Caffeine is a diuretic: among its negative effects it causes you to urinate more than the volume of water contained in the beverage. Similarly, alcohol dehydrates, particularly the brain, and is therefore a depressant. It also has the effect of dilating the blood vessels of the vocal cords, reducing the throat's necessary sensitivity and suppressing the immune system.

Water management

Dr Batmanghelidj suggests drinking a minimum intake of eight glasses (about four pints) of water spaced throughout the day.[14.9] He cautions against both over-hydration and the loss of salt. He writes that salt is a strong antihistamine and an anti-stress element for the body; it helps to clear catarrh and sinus and lung congestion, avert dry coughs and maintain muscle tone. He recommends using unrefined sea salt, which has not been stripped of its many healthy mineral elements and does not contain additives such as aluminium silicate, poisonous to our nervous system. He cautions against excessive salt intake, noting that the critical ratio of the body's salt and water needs in conjunction with other minerals must be carefully observed, and that one should always drink enough water to wash excess salt out of the body.

Dryness and the mucous membrane

Although research into the significance of the membranous lining of the respiratory tract and its condition is still in its early stages, singers have always been aware of it, mainly through feelings of dryness and the seeming capriciousness of mucous. The mucous membrane of the vocal tract is highly sensitive – the tiniest particle of dust triggers a cough. It doesn't like dust and dryness any more than a car engine. In fact, the smooth mechanical efficiency of our voice – its precision and athleticism – is dependent on the healthy condition of the mucous membrane.

The larynx, apart from the folds themselves, is duly supplied with numerous mucous glands whose task it is to keep the area appropriately moist. Where the folds meet, however, the membrane is of a minutely scaly tissue to reduce the effects of friction. One can easily imagine problems arising either from excessive vocal fold friction or when the secretions of mucous are in too short supply, or too thick or sticky. A combination of friction (from inefficient closure) and dryness is probably a major cause of vocal ill-health. The better a voice works 'mechanically' the more easily it is able to withstand thickened mucous.

Dryness is one of the worst conditions for the voice. There are many factors that can contribute to this anti-singing condition (apart from recognised pathologies and illness such as inflammation, oedema, infections, blood blisters and nodules, which must be considered abnormal). Dryness can be caused by poor air-conditioning or lack of humidity in the air, and by atmospheric pollutants, which can also cause a tight chest, effortful breathing and irritation in the eyes, nose and throat. Smoky and dusty atmospheres should be avoided, as should irritating products such as toxic paint or household detergents. Medications can also contain drying substances. Chapter 40 of *Professional Voice*[14.10] covers antibiotics, antihistamines, inhalants, sleeping pills, analgesics and so on. There are other informative chapters on conditions such as reflux, allergies, pollution and endocrine dysfunction. The writers of Chapter 19, R. T. Sataloff, K. A. Emerich and C. A. Hoover, state 'The human voice is extremely sensitive to endocrinologic changes. Many of the voice changes are caused by alterations of fluid content beneath the vocal fold mucosa.'[14.11]

Immunity

The average singer breathes more deeply than the non-singer. In spite of obvious advantages to her health, this may make her more vulnerable to germs and viruses. I believe that a properly lubricated mucous lining makes her less likely to succumb to these hazards. Germs are easily spread by hand, and public surfaces, such as handles, shopping trolleys and money, are potential germ conveyers, as are piano keys and other peoples' hands! Immune system maintenance is imperative for singers, who generally cannot even afford a common cold. Breathing through the nose, a vitally important filter and air warmer, is sensible indoors as well as outdoors in any weather. Sunshine is good and necessary, but singers should beware of drying out.

Breathing health

Breathing well is vitally important to our general health. The oxygen we inhale is conveyed to our muscles through our blood (or more accurately by the haemoglobin in its red cells), and pumped around our body by the heart. Carbon dioxide (a waste product) needs to be got rid of. If we breathe out weakly, this waste is incompletely evacuated and the natural replenishment of oxygen is correspondingly superficial.

Muscles that have to be more active require more oxygen, which demands increased cardiac activity both in terms of heart beat rate and the amount of blood that each beat pumps (stroke volume). The heart calls on the lungs to supply the oxygen, and so our breathing system works harder. Stronger exhalation meets the increased demand, improving circulation while exercising the crucial inbreathing reflex. The full extent of exhalation demands participation of postural muscles which keep us both 'proud' and grounded through their back-lengthening work.

Sleep

Our body (our voice) needs rest and recuperation to be fully alive in every pore for training sessions, rehearsals and performances. I believe that diet, digestion, regular eating patterns and exercise all play a part in the quest for a good night's sleep. Singers with 'crazy' schedules must work out how to cope in non-conducive circumstances, but there is no doubt that a tired body makes heavy weather of the 'totally involving' activity of singing. Tiredness can also exacerbate the negative effects of dehydration.

Stress and worry

Stress, which most of us suffer from to some degree, manifests itself physically in tension, hyperactivity and various illnesses. Tense, we are more likely to make mistakes or set ourselves up for disappointment or negative outcomes. Stress and worry impede our performance, making us over-deliberate or act impulsively, erratically or not at all. When stressed in performance, we are bound to try too hard because tension affects the working of our muscles.

Some people manage over time to deal with stress in a rational way, and I believe that, since many worries are imaginary, 'talking to oneself' must be part of the 'treatment'. If, for example, you have prepared your music thoroughly for an audition, thought of all the questions the panel might ask, kept fit, and visualised walking confidently onto the stage and away with the prize, you have taken all the positive measures you can. Remaining factors are unknowns. You cannot dictate conditions, you cannot make the adjudicators choose you or even like you. Worrying without cause will only spoil your performance, and may reinforce your feelings of unworthiness.

There are many ways of relieving stress, from meditation to having a good laugh. Exercise, depending on its strength and duration, can leave one invigorated or needing rest. Either way tension is released. My best advice is to practise living the moment, savouring and taking care of exactly what you are doing – the life you are breathing. It is after all the only rational thing you can do, as well as the best way of achieving something of quality, which will by definition be paving the way for what comes next. Really concentrating on what you are doing is a kind of meditation which tends to banish worry and be highly productive.

Talking health

The normal speaking voice is notable for its lack of harmonics, the 'thinning' process of the cords being generally weak. To be healthy and colourful (musical) it must have its mean pitch from which it can range and inflect widely, a concentrated but flexible core, neither too hard, tight or sharp, nor too soft, breathy or ill-defined. For this the speaker must be appropriately animated. The cultivated 'singer's voice' is in reality quite inflexible, and just as likely to end up in the voice clinic as a normal, overworked but under-supported voice. As we have seen, it is the extrinsic muscles of the larynx which, in actively suspending it, release and support it for healthy vocalisation. It may be a grave mistake, however, to think that this is their only purpose. Thinking in this narrow, reductive way perpetuates singing's exclusivity, and overlooks the other benefits of a 'well-supported' larynx.

The jungle of muscles in the throat have many different tasks to perform. Between them they support and articulate the head and aid movements of the jaw, yawning, swallowing, sneezing and so on. Some of these muscles are involved with singing and therefore with breathing. Logically, where there is mutual muscular dependency, any weakness will render the throat as a whole less efficient and more vulnerable to stress or strain. A fully-functioning larynx with which the body – its posture and breathing – is so intimately involved must benefit the many other functions of the throat.

—

To sum up, bearing in mind that everything we do affects our ability to sing, singers must pay due attention to physical well being (what we eat and drink is materially what we are), attitude to life and living (how we think can have a positive or negative influence) and life style (what we do can benefit or inhibit our performance).

PART II

Chapter 15

Gurus or Guides? (Teaching)

> *Nobody ever teaches anyone to teach, so let's not pretend. You learn how to teach by opening up, by questioning, by doubting, by exploring, by rebelling. You learn how to teach by learning how to learn.*
>
> Eloise Ristad [15.1]

Introduction

As a young student of singing I became increasingly intrigued by the teaching process; not so much by how a voice worked, but by what it could be encouraged to do and what that seemed to mean. One day during a lesson with Yvonne Rodd-Marling [15.2] I asked her to teach me how to teach. She replied 'Just listen, ducky, use your ears'. I was disappointed! The implication seemed to be that if I wanted to teach I was on my own, and at that point I had little idea what my ears could do for me. However, I was so in awe of this extraordinary teacher that in spite of my misgivings I began listening objectively there and then.

For many years now, I've known that the advice of Rodd-Marling was the best she could have given me. In effect, she was saying that if I wanted to find the way to teach I could do this only by exploring with my own ears. And I am aware that as long as I'm listening and hearing I am still learning, still on the right path.

Among some of my old notes I recently found: 'Whether a teacher is on the right track or not would seem to depend on his definition of singing'. The implication is that you must *define* singing before you can teach it or that in her own eyes a teacher is working correctly if she's achieving the results she expects. I rejected this view, and eventually found a response to this widespread but erroneous idea by stealth. I resurrect it here to make a point. Defining singing (and there are many different definitions) is more likely to cause problems than aid solutions. As far as I can make out, definitions are always based on what the voice can *do*, rather than what it *is*, on what can be seen 'on the surface' (such as good coloratura), or on what we expect or need of it.

As illustrated in Part I, it is clear that we all have a singing voice, and are therefore designed to sing. It is also clear that humans in general have a deep-seated desire to sing. The singing voice, however, has suffered from varying degrees of neglect and abuse, thus frustrating this desire to some extent. There seem to be two solutions to this problem. The first is simulating or 'faking' a singing voice, by learning how to make what you imagine to be a singing sound, or by devising ways of satisfying the requirements of the music you are singing. This usually amounts to coping with difficulties without eliminating them, and of skilfully manipulating the voice to make different sounds as desired – a mind-directed process, in which the voice remains under conscious control.

A far more effective solution is that of releasing the voice. This is a sense-directed process which aims to restore something we possess by nature, enabling it 'to be'. It acknowledges that the singer and his voice are one and the same, and that in enabling the voice to achieve its potential we are enabling the person to fully express himself in sound. It's understood that the various attributes we normally ascribe to singing (skills, if you like) are inherent to the instrument, and that matters of colour and character, tessitura and dynamic range are unique to each individual. In considering the broader human implications of our work we must guard against conflict between what we want or expect from a voice and what it is, or might become.

Responsibility

I remember once reading that a teacher affects eternity, and can never tell where his influence stops. This could be applied to other professions and is true of personal relationships. The reason that a teacher can have such an influence 'on eternity' is that he or she is supposed to have the answers, and to convey 'the truth'. Teachers (like priests, gurus, and even parents) are often invested with undue authority, especially when their charges relinquish their own responsibility.

The truths that teachers convey to their pupils, however true or false they may be in fact, are passed on, elaborated or distorted for new audiences or new situations, involving and affecting others as they are passed on. I remember several occasions on which 'answers' said to be my own came back to me in a revised version! The one-to-one teaching relationship is a particularly powerful and influential one, in that it is often assumed that the pupil is dependent on (if not in awe of) the teacher, who is going to make him what he wants to be. Unfortunately, much that the teacher says is turned into an 'answer'. One can only hope the result of the 'doing' in lessons is more productive than what one says! Singing teachers are not repositories of the truth – the best we can hope for is to be conscientious seekers *after* the truth, which will, as this book illustrates, be different for each individual singer. There are dangers on both sides in not being aware of the negative effects of the 'all-powerful teacher', whether he or she feels infallible or not.

In *Feet of Clay – A Study of Gurus*, the psychiatrist Anthony Storr writes:

> Psychotherapists are familiar with the occurrence of transference, a phenomenon first described by Freud as the process by which a patient attributes to his analyst attitudes and ideas that derive from previous authority figures in his life, especially from his parents. Later, the term became extended to include the patient's total emotional attitude towards the analyst. Freud at first regarded transference with distaste. He wanted psychoanalysis to be an impersonal quest for truth in which the relationship between patient and analyst was entirely professional and objective rather than personal. The role he wanted to assume was that of a mountain guide. Instead, he found that his patients made him into an idealised lover, a father figure or a saviour.[15.3]

Some teachers enjoy their pupils' dependency and hero-worship, and take advantage of it. Young pupils in particular can be gullible. Storr reminds us:

> The very young perceive that their parents know more about life's problems than they do, and it may take years for a child to realise that his parents are not omniscient but fallible. A lingering hope that somewhere there is someone who *knows* persists in the recesses of the minds of most of us, which manifests itself more obviously when people are distressed or ill.[15.4]

The most dangerous pupil-teacher relationship is the one in which the pupil has a strong feeling of dependency and the teacher has a strong sense of his importance and professes to have the answers. Each is 'seduced' by the other into an unhealthy relationship in which the pupil entrusts himself, heart and soul, to someone who is only really interested in his image and position, and perhaps wants to be admired. The pupil relinquishes responsibility and the teacher assumes control. While this is an extreme situation, the ability to keep your mind open to new discoveries while assuring your pupil that you know what you're doing can be difficult.

Authority is often difficult to question, and being in an authoritative position as a teacher, you can feel secure and inscrutable. This position, however, is often inflexible and dangerous; the teacher tends to stick to what he knows and what he wants and therefore is unable to see or hear, let alone fully attend to a pupil's needs. This rigidity limits both teacher and pupil. Teachers should beware not only of holding up a pupil's progress through lack of knowledge or an unwillingness to explore, but of transferring their own faults – vocal problems or mannerisms, prejudices or even performance nerves – to the pupil who may copy if the teacher isn't sufficiently vigilant. Admitting that you 'don't know' may threaten your credibility and position, but it makes you a better teacher, enabling you to be more searching and inventive. It frees you to say 'let's explore, let's try this', and to profit from the team-nature of the work.

From punctuality to fitness to do the job, teachers must set an example if they want good results. In teaching, neglect on the part of the teacher is as counterproductive as the imposition of her ideas. Great singers don't necessarily make great teachers: they are inclined to think that the way they sing is the way their pupils should sing (and their pupil may agree) and they often have no idea how their own voices work or reached their present condition. It may be even worse if a singer has turned to teaching due to vocal problems, especially if the reasons for these are inexplicable. The idea that a teacher should be a better singer than his pupils is erroneous, because it assumes that what the teacher can do himself is more valuable than what he can get from his students.

The philosopher Martin Buber wrote:

> There are two kinds of therapists, one who knows more or less consciously the kind of interpretation of dreams he will get; and the other, the psychologist who does not know. I am entirely on the side of the latter, who does not want something precise. He is ready to receive what he will receive. He cannot know what method he will use beforehand. He is, so to speak, in the hands of his patient.[15.5]

Instead of assuming superiority to our pupils we must work alongside them, with the work a partnership and shared responsibility.

The teacher's voice

A teacher must at all times take care not to damage her voice. It is not sufficient simply to be careful not to overdo it. How does a teacher keep in training? One obvious answer is to have lessons or periodic 'check-ups' with a respected colleague. If this isn't possible, I consider it both desirable and legitimate to use the lessons you are giving to check on and exercise your own voice. In demonstrating for your pupil, you can beneficially make more or less of this or that sound (vocal event) according to your own voice's condition or needs. Doing this alongside a pupil helps to concentrate your own self-hearing.

Role playing

It may be useful to consider whether role-playing contributes something positive or negative to our work. Adopting such roles as Mother Hen, Prima Donna (retired) or Svengali can block our true ability and that of our pupils, who may nevertheless be somehow attracted to them. It may be hard not to wield power when it's in our hands, not to let our caring nature stifle our charges, not to preach about something we believe strongly, not to become 'smitten' with an attractive, enthusiastic pupil who is pouring her heart out, hard not to go on promoting oneself after the curtain has come down. The paradox is that in teaching singers we're dealing with the interface between the ego (as in self-idealisation, or self-interest) and the self (as in true individuality or essence). While a singer must 'play' a role on stage, singing is a direct expression of body and soul, and our deepest communication with others. It is this rather than the 'show' of performing that gives singing its potency.

Talent and trainability

There is all the difference between a voice that seems to 'have something' and one that's trainable, given time and resources, to a professional standard and all that that entails. In so far as the process of teaching singing is the freeing of the person, not just transforming him into a marketable product, teachers have a huge responsibility. Do we or do we not embark with a young person on this exciting but exposing and potentially hazardous human journey? Does the singer have a reasonable chance of success, of staying the course? How complicated are the singer's non-vocal issues? Is the singer fit or prepared to get fit? Are we equipped to deal with the case? More searching questions need to be asked and answered before a decision is made. A trial period of work gives both teacher and pupil the chance to assess what can be reasonably expected of one another.

To select a would-be professional who is not prepared to get fit and put in a lot of disciplined hard work is a waste of time, money, energy and expertise. On the other hand, much can be achieved by a singer demonstrating modest talent but prepared to work hard enough for long enough. Life out there is tough, but making the grade doesn't have to be a question of 'the survival of the fittest'. A good attitude and good energy, in addition to vocal and musical talent, may be sufficient to begin with. If a would-be singer understands what's involved, he has a much better basis upon which to decide how seriously he's prepared to take the work. I'm convinced that being realistic from the start prevents more heartbreak than it causes. Individual teachers and selection panels can be wrong about prospects and I advise would-be professional singers to consult several teachers before deciding on their futures.

Relationship and trust

The desires to teach and learn are closely related to the desire to connect with others. In order that the process is effective, teacher and pupil must acknowledge the interdependent nature of their connection through their work. This can be healthy and productive providing they are both subscribing to the same process: enabling, integrating and growing. The teacher has knowledge and skills of which the pupil has need – a context for teamwork. The pupil helps the teacher in his work, as much as the teacher shows the pupil how to help herself. What counts is the integrity of this dynamic from moment to moment and situation to situation. In this sense integrity is process.

Since the singing voice literally *is* the person embodying it, it confronts people with themselves. The teacher is in effect 'delving into' his or her pupil, and it's impossible therefore to avoid a 'personal' element in the relationship. As with the doctor-patient relationship, there needs to be mutual respect and trust, the patient trusting the doctor and the doctor being trustworthy. The teacher is normally older, and more experienced both in the teaching field and in life. On some level the singer will realise that singing is an 'opening-up process', and therefore, the onus is on the teacher to inspire confidence in the pupil – in what he does rather than simply in what he says – so that trust can follow. A good teacher sincerely desires the good of the pupil. This makes work more effective, (as it does, remarkably, in healing) helping the teacher to discriminate and be concerned with what *is* genuinely good and what is not.

In my experience pupils are sensitive to the sincerity or otherwise of teachers' concern and respond accordingly. While genuine mutual admiration can be highly productive, unrealistic expectations, either way, can be disappointing and damaging. While a little teasing may lighten the work, behaviour in which a hidden agenda is 'felt' will easily put even the least sensitive pupil on the defensive. Conscious or not, this can seriously impede or complicate the work. Pupils can sometimes invest a teacher's genuine kindness or attentiveness with bad intent. In any case, teachers need not only to be sensitive to their pupils' reactions in general, but to monitor their own conduct and feelings, observing their effect (positive as well as negative) on pupils and the work as it progresses. A teacher may need to modify her behaviour to make a pupil who tends to be guarded feel secure. On the positive side, a teacher's enthusiasm about her subject can communicate itself to her pupil, uniting the two in their quest for 'the truth which transcends personal considerations'.[15.6] (*Note:* In normal circumstances, I do not consider it necessary or desirable for teachers to touch or physically manipulate pupils in lessons. This practice inevitably encourages physical manipulation on the part of the pupil, [who will already be overly inclined to go by how things feel], and draws attention to things which are better off without conscious control.)

Communication

Genuine intentions can move the work on much more quickly than mere purpose; they encourage pupils' wholehearted co-operation simply through the quality of energy that they transmit, and can make the pupil actively want to get more involved. This is particularly important with aspects of the work which 'look strange'. For good progress, the teacher-pupil relationship has to be open and honest. By this I don't mean free-for-all question and answer sessions, but rather that both teacher and pupil should feel free to say what needs to be said and to ask and answer questions appropriately. This is a state to strive for, in which I believe the teacher must lead the way, being focused but clear about what she wants and what in each case it's necessary to do or avoid. The teacher mustn't be afraid to state the obvious, improvise, or try novel ways to approach difficulties. The sooner she overcomes her own reticence or lack of confidence, the sooner her pupil will feel confident to open up and take the initiative. Child-like playfulness is facilitating because it is communicative, creative and unifying. Teacher and pupil need to invoke this spirit and nourish it in safe surroundings.

Criticism

In an interview with *What is Enlightenment* magazine, the Romanian poet and social critic Andrei Codrescu states, 'The potential of human being is in commenting, translating, judging the world.... An unwillingness to judge only reflects a lack of courage and it's the disease of a politically correct culture that is afraid to offend.'[15.7] The word 'judgement' has a sense of condemnation and finality about it and many are afraid to either be judged or to be thought judgemental. Discrimination can be lost to prejudice. In fact 'judgement' and 'discrimination' have useful meanings. To 'discriminate' is to make or see a distinction or to differentiate, while 'judgement' means the critical faculty, discernment, good sense, and many other invaluable things like perception, insight, acuity, and plain common sense. It's difficult to see how any teacher can get along without these ways of evaluating, moment by moment, specifically and generally, their pupils' work.

Pupils need helpful feedback, not merely the terms 'good' and 'bad'. Criticism must have a beneficial point and always be constructive. 'Good' may only be useful if a pupil *knows* that what she has done was good. Negative criticism is at the very least disheartening, and can be destructive, especially when delivered in public. Remember that singing is a self-validation of a kind; there's no other field (except perhaps acting) in which criticism of what one does is so personal by implication.

All criticism should be appropriately pitched. This doesn't mean dependent on whether the pupil is just starting lessons, or is a student or professional, choir member or soloist, but rather on what you perceive to be the individual's capacity for work and improvement. Of course if the question is 'Am I ready for music college, or to audition for a major opera house?' a different type of assessment is being asked for. Criticism should not be based on the progress of others. If a student is always being measured up against his more advanced peers, or against where the teacher thinks he *should* be with his voice, he is likely to be sorely disheartened.

Given that the only true yardstick in criticism is the individual's capacity, the type of criticism proffered depends on at least four criteria in order to be constructive:

1. **Its purpose**. This may be to encourage, as in 'can you hear how much better that is?' Or, when the pupil is in doubt, 'That was great, because . . . can you see that?' Criticism can be used to deter (in the sense of, 'that particular way wasn't helpful, can you try . . . ?'), or to make the pupil aware of something going on, or something that needs to be attended to. It can be used as a reminder, or in a review of work done.

2. **Its specific basis or context**. Criticism should refer to something in particular, recognisable by the student and giving him a point of reference or something to build on. For example if the focus of an exercise was on how it was approached (spontaneously, rhythmically etc), and there was improvement but with one or two notes out of tune, there is no point criticising intonation. The student needs to know that what he did fulfilled what was required by the exercise. The success of criticism here depends on how awake the pupil is to the experience he's having. If he didn't notice, in this case, that his approach to the exercise was more rhythmical than before, the teacher should point this out, encouraging the pupil to be his own detective and to pay closer attention.

3. **Being realistic.** It's very easy to criticise, especially over superficial matters, which in fact are rarely important. Criticism is only valid if the pupil can reasonably be expected to have done better. While a conscientious teacher will always try to present his pupil with suitable challenges, even the best teachers don't always gauge this well. Criticising for the sake of it is unhelpful and can be dispiriting. Even the simplest exercise may have various facets and whether a pupil can fairly be criticised on more than one or two at a time must be carefully judged. Feedback must be clear and uncomplicated, and something that the pupil can relate to.

4. **The spirit in which criticism is delivered.** Sarcasm and mockery can have very negative effects, as can boredom or exasperation. The delivery of criticism (choice of words and tone of voice) can have as much bearing on the process as the idea behind it. Genuine pleasure may be more powerful than matter-of-fact acknowledgement of a good job, but the latter may be more appropriate. What is important is that the pupil feels encouraged to go forward. Criticism must therefore be apposite and empowering. Giving value to what we are doing and what is taking place must be our constant concern since it is this that keeps us and our pupils alert and on the right track.

Evaluation

One of our major challenges as teachers is to retain and balance both the analysis of a voice and its synthesis in our minds. We have to work on the different parts, keeping the whole as it stands in mind, then gauge to what extent the parts contribute to the whole, and how the whole may develop with the increasing *ability* of the parts to contribute or merge. As Margaret Wheatley puts it, 'we enquire into the part as we hold the recognition that it is participating in a whole system. We hold our attention at two levels simultaneously'.[15.8] This is primarily a matter of aural perspective, being receptive to the individual's 'vocal landscape', although to better analyse the sound we may at first need to visualise the anatomy of what we are hearing. We cannot assume the validity of the measurement or quality of one aspect unless we value it in the context of the whole – the part as contributor and the whole dependent.

As teachers we must develop ways of illuminating 'the picture' we hear, and of scrutinising it from all angles so that we don't miss important cues or leads, or base evaluation on false criteria. Exciting creative departures and possibilities are often chanced upon regardless of our intentions. As my friend Richard Hames once put it, 'constant process of change needs appropriate response to changing conditions'.[15.9] The spiral (a symbol of continuity and returning to the centre) is an aid to maintaining perspective. There are others:

1. **Bringing the long-term to the short-term view.** We remind ourselves that our objective is not a short-term expedient but part of a longer-term process, the success of which depends nevertheless on taking the right steps moment by moment.
2. **Suspending judgement.** This means not evaluating something until we have gone far enough with it. There's only so much we can say about progress without repetition. Some paths are convoluted, and progress is made for a while, seems to stop, then re-emerges more advanced. Premature judgement can confuse pupils.
3. **Reconciling what we do with what we hear.** Thus we avoid doing exercises for the sake of it.
4. **Not blocking progress with expectations.** A pupil may not give an exercise her best unconsidered shot if a specific outcome is expected. Evaluation must not be based on an agenda. Similarly, we must distinguish truth and reality from the official line. There are *no absolutes* in singing, so outcomes should not be measured against what is officially considered 'right', any more than we should adopt an accepted 'way'. Beware of evaluating voices on the basis of what you're used to hearing. If we expect anything, it should be something unexpected, something unique to the singer.

5. **The more we approximate to truth, the more we should guard against error.** When we think something is 'right' we should test it out more rigorously.

Ringing true

Descriptions of sounds are not absolutes. There are many varieties of the sound described as *falsetto*, for example, and even more of head voice, which is why singing teachers must determine what is *behind* every sound they hear. Even then they must appreciate that each sound belongs exclusively to the person who makes it. Aural perception depends on many factors, among them knowledge of sub-structure and vocal attributes in general, the ability to distinguish an individual's sounds from known vocal sounds in general, and the ability to compare an individual's own sounds. Identifying a person's sound is ultimately a question of *recognition* – not knowing 'how' or 'what' but recognising that a sound 'rings true' for a particular person. I would say that in general people don't trust their intuition or instinct enough. A teacher has to learn to guard against preference or prejudice. Seeking support for prejudice seems to be a strong human trait. The non-cerebral discernment of a singer's sound has always been an invaluable source of knowledge, and the development of technology only reinforces my belief that singing would be impossible without this non-cerebral aural 'knowledge'. Young children react to tone of voice, and know when it's 'put on' – this is the 'intuitive response' to true and false that I'm talking about. All sounds can and must be checked and balanced but when a voice rings true there is nothing more to be said.

Specialist knowledge

In our aural task, we are reliant on the immediate detection of minute variations in muscle tension, and being able to act instantly, whether this means urging the singer on, changing course or stopping abruptly. We must gain exact aural knowledge (generally and with each pupil) and become sensitive to both ambiguity and surprise. Our ears are the only competent organ for vocal analysis and synthesis: apart from being able to distinguish thousands of sounds per se, they are quicker to observe vocal tone than the mind is to analyse it, and have a deep vein of natural knowledge, if only we can tap into it. It is the teacher's job and his skill to be 'the singer's ear' until such time as she recognises her own voice from the inside, until her aural picture comes back into focus.

Fresh beginnings

Apart from any relaxing or general physical warming up exercises a teacher might consider necessary, the beginning of the process – every session – is the emission of a sound, a unique event which should be heard and valued as such. On the surface at least, it's relatively easy to hear what is 'wrong' with a voice, but a teacher's ears must always penetrate more deeply, searching for what is genuinely *good*. It is only then that we have a chance of recovering the whole.

It can take months to know a pupil and his voice (strengths, weaknesses, inclinations and idiosyncrasies), and to know what effect the condition of his voice has on him and what he does to avoid or make up for difficulties. The process must be continually re-informed by the developing teacher-pupil relationship. Even then a voice can spring surprises, either positive or negative.

Because the voice and the emotions are so closely related, it's possible to approach an uncomfortable emotional area in a singer prematurely. He may be able to say 'That's too close for comfort!' But singers who are not aware of what is going on emotionally often 'back away' or 'close down' instinctively. Both cases signal something new to work through. Energy that inspires and feeds the process is relatively easy to attain at the outset of training, but we need to find ways of recreating and sustaining it. Teaching singing is not so much a matter of method as of recognition, understanding and resource.

The 'feel good factor'

Feeling good or satisfied is not an aim but a consequence of a process which may at times be emotionally uncomfortable and physically strenuous, invariably taxing our concentration and often our patience. Delving to deeper vocal and emotional dimensions can be painful as well as liberating. While effort in itself adds no value to achievement, training is invariably hard work until the voice is fully liberated.

Barriers to progress

Inevitably, there are many things that get in the way of progress. Lack of trust, usually but not always on the part of the pupil, can be a particularly difficult one to overcome, and not always easy to detect because of its visceral nature. Some other obstacles that can impede the teaching process follow.

1. Information

Most singers want to know how their voice works. It's often naively assumed that if only you knew how your voice worked you could make it do what you wanted. Some pupils want to reassure themselves that the teacher knows what she's doing, while teachers may want to convince them that they do. Teaching singing is a highly active process. Explanations may result in 'doing', but rarely in freeing, because they encourage neocortical control. So long as a pupil does not 'know', he can be guided to direct his attention appropriately. Deliberate withholding of information should be explained if necessary.

The time for explanations and information is 'after the event' when the pupil is already doing something well. Explaining before a pupil has *heard* can seriously hold up progress. Teachers should avoid passing on 'received information', or explaining something to which the singer can't relate. Remember, we're enabling the singer to sing by drawing out his or her own unique voice. *It* informs *us* as it re-emerges. A pupil must be helped to recognise her voice and to own it.

2. Terminology

Terms are not absolutes. Each singer's 'head voice', for example, sounds different from the next. Even many terms used in this book, such as 'antagonists', 'structure' and 'forces', can be misconstrued as something negative. Voice terminology, so as not to be misleading, should accord with what's heard by both parties. Any term might be used to 'label' a sound, or a type of sound, so long as both teacher and pupil hear it for what it is. If someone asks 'Is this my chest voice?' I'm inclined to say, 'That's the sound you're making; what does it sound like to you?' I tend to use terms like 'chest voice' only when my pupil perceives it as sound rather than a position or action.

3. Exercises

I have always maintained that once an exercise is written down it's redundant. Individuals' needs are different, and continuously changing, so individually changing treatment is required. I have often observed students at workshops writing down exercises – their energy and attention is diverted, and the exercise's value missed because the progression in real sound has not been registered. Teachers sometimes ask me what exercise I would use for a particular problem, which they then describe. I cannot prescribe an exercise without hearing the sound of the individual singer and his individual problem. There would then be the question of investigating the problem's roots. Exercises, like terminology and the labelling of singers, are sometimes inappropriately used for lack of relevant consideration and thorough investigation.

A pupil occasionally catches me out by asking why we did a particular exercise. If it was the culmination of a series of vocal manoeuvres I'm reluctant to explain, because the exercise was as much the manoeuvring as it was the final outcome. The correct answer to the pupil's question lies in his experience – it was so as to experience *that*.

There's rarely any point in insisting on an exercise that fails to 'strike' on second attempt. Pupils sometimes ask me to repeat an exercise. I rarely do so because it was done in and for the moment. A repeated exercise is usually approached too thoughtfully and hesitantly. Instead of explaining or repeating exercises, pupils should be encouraged to be more aware of the experience they are having in all its aspects, especially the aural.

4. Practice

Singers in general have almost invariably lost 'aural touch' with their voices and cannot exercise on their own without resorting to the physical sensations and thought processes which prevent spontaneity. It is of course unrealistic to expect a singer to remain mute between lessons. I suggest two compromises. The first is experimenting with some of the exercises experienced in lessons, with the proviso that they are approached in the way that all exercises should be (see Chapter 13). These exercises should only last a few seconds at a time so as not to encourage imbalance and should not be practised. The second compromise is singing spontaneously (when the pupil feels like it), not in a studied fashion, not forcefully or insistently, but with animation and joy. Emotionally expressive music with a lively pulse should be chosen. Certain vocal elements can be practised once sufficient progress has been made and the teacher is satisfied that the pupil understands their purpose and knows what to listen for.

Old habits can be reinforced by singing, and occasionally it's necessary for a pupil to stop singing for a while. François Couperin gave half-hour lessons every day, then locked the harpsichord, so as to minimise negative effects of undirected or purposeless practice. It seems strange that singers think they can improve their instrument and their use of it by singing for hours at a stretch with no real idea of what they're doing. In the absence of Papageno's padlock, a singer must decide for himself how disciplined he is prepared to be in order to attain his goal!

5. Technical aids

I am old enough to be someone who prefers to use a pen but I use my computer because it is quicker and more efficient. Writing being manual and external, it makes sense to improve the activity by external means. In singing, nothing can take the place of the ear, which is a natural built-in monitor and therefore part of the voice itself. The ear does not need anything to do its job for it, and so 'external aid' is a contradiction in terms. Even 'aids' that come from the body itself – attempts at physical control – make an awkward, self-conscious job of singing because they're not integrated. The great value of training is in the restoration of the ear-voice relationship. This ear-voice dynamic will only recognise itself by experiencing itself. All forms of external physical aid are thought-directed, distracting and delaying. We think, then interpret, then act. This almost always leads in a different direction from the ears, which otherwise hear and adjust simultaneously.

Singers are often surprised when they hear their recorded voice. This can be lack of recognition. But however sophisticated the recording equipment used, the result is never the naked truth. Neither the singer nor the listener are sharing the moment or experience, and the sound is a 'clothed' or even 'doctored' version. Recordings can be useful musical and interpretative aids as long as they are treated as such. I have rarely allowed a pupil to record a training session because, where sound is everything, the recording quality is often less than good, and, more importantly, the purpose of training is to induce 'vocal events' heard as sounds by teacher (who can give feedback) and pupil (whose inner ear is in the process of being retrained) in the same moment. In private the pupil is bound to imitate the recorded sound (useless in itself) and base subsequent adjustments on false value-judgements, thoughtfully and manipulatively. What is done spontaneously in the lesson is part of a progressive process, which by definition cannot be reproduced. Similarly, looking at some technological monitor, even in real time, discourages hearing, and encourages delayed interpretation, so that by the time the singer has registered what was going on, what needed correcting has passed.

6. Specialisation

As teachers of singing we're not responsible merely for a musical product, we are dealing with people. A psychologist would not dream of encouraging a particular side of a patient's character at the expense of his other strengths and qualities. All free voices have agility, carrying power, dramatic ability, expressive qualities and so on, and thus, encouraging a voice to become a 'type' because of a particular strength or tone colour preference, is as irresponsible as it is erroneous.

7. Technique

People describe singers as having 'good technique'. This could mean two radically different things: artifice, with everything achieved through the ingenuity of the singer rather than through vocal merit, or a high degree of vocal freedom, with ease and natural skill, and probably beauty, as well as artistic merit.

Frederick Husler and Yvonne Rodd-Marling (page 112) are very clear and succinct about this subject.

> Organic being has no capacity for living 'technically'; to impose technical measures upon it invariably signifies the presence of some alien force. Technique, in short, is not a physiological term. Of course the singer, and especially the voice trainer, cannot altogether dispense with so-called technique, if the problems involved in singing are to be dealt with successfully. The latter must have recourse to 'technical' practices to unlock the organ, while the singer is forced to employ them because what he has to perform often exceeds the present capacity of his vocal organ. 'Technique', in other words, is a useful tool but nothing more; a crutch, as it were, to help the unfinished or ungifted singer. [15.10]

PART II

Chapter 16

A Clean Slate (Learning)

This is the setting out.
The leaving of everything behind.
Leaving the social milieu. The preconceptions. The definitions. The language.
The narrowed field of vision. The expectations.
No longer expecting relationships, memories, words or letters to mean
what they used to mean. To be, in a word: Open.

Martin Buber[16.1]

Learning to sing is an exciting and invigorating adventure. It can also be frustrating and frightening, because it's a journey into the unknown. No one knows how long it will take, how easy or difficult it will be, what or who you might meet on the way, or the eventual outcome. Your teacher is your guide, no more, no less. He or she has knowledge of the singing voice, and has made many similar journeys. Your journey, however, is a new one for him or her. And like mountaineering, this is going to depend on team work. A teacher can't do anything without the pupil's co-operation. As pupil you must do your share of 'reading the map', looking out for signposts and landmarks, and being responsible for your own kit and fitness. Your guide is not paid to carry you! As my colleague Lorna Marshall once put it: you are in effect ticket, map, vehicle and fuel.[16.2]

If we understand to what extent the voice is the self we'll also understand how personal this journey is. When you learn, you are learning about yourself, and your relationship with yourself is logically the most important there is. Training is an opening up and illuminating process. Your singing voice is your 'whole self' reflected in sound. This is what you are discovering in learning to sing, and why it can be both exhilaratingly exciting and fraught with frustrations.

You don't need lessons to *give* you a voice but to release the one you have. You don't even need someone to show you how to sing, because a liberated instrument responds spontaneously to the *desire* to sing – on some level you already know how. I'm not talking about learning how to sing music by Purcell, Mahler or Britten.

Learning music by a particular composer is a question of style for which you do need knowledge. I'm talking about vocalising, which human beings are designed to do. Students often think they need to learn how to sing higher or lower, rapidly, strongly, with clear diction, a long line, and so on. It may be a consolation to know that these skills are natural attributes of a fully liberated voice.

Liberating your natural singing voice can take time because of your particular vocal history: what has happened to your voice through childhood until the moment of taking singing seriously. What I've called the 'civilising process' might otherwise be called 'conditioning'. This means that the ways we think, feel or express ourselves, and the way we conduct ourselves physically are influenced by our upbringing, education and the society in which we live. Much of this process has a negative effect on our ability to function fully and freely as individuals and is reflected in our vocalising. Instead of developing according to their potential, voices get out of alignment, or 'close down', so that when we want to sing they respond weakly, incompletely or effortfully. Those who survive this process relatively unscathed we call 'natural' singers, because their voices have fulfilled their potential to an unusual extent. Describing someone as having 'no voice' simply means that it's well and truly locked away.

To regain your true singing voice (rather than merely making the most of its current poor condition) it has to undergo a process of coming into renewed existence. To do this effectively, you and your teacher must tackle the difficulties at source, thereby encouraging your voice to find its full extent and depth. Vocal regeneration can take longer and require more dedication than acquiring a serviceable technique, but ultimately it is far more satisfying simply because it's genuine – it 'fits'. Rather than 'patching up' your voice or 'pasting on' effects, you are restoring it to its natural condition, liberating your musical potential while satisfying your desire to express yourself fully and directly. A major reason why it can be difficult to find your voice is the confusion that often arises between who we appear to be and who we really are, between the identity or image we have acquired or cultivated and our true, often unfamiliar identity as revealed in our singing voice. Finding your true voice may entail the unravelling or unmasking of unfamiliar parts of yourself, and a radical change in how you are seen by yourself and others.

If a singer thinks that her voice is fine but feels its limitations (in range, dynamics, flexibility and so on) and the need to 'make up for them', it's because she has got used to a way of singing which simply prevents her voice's skills emerging. Some 'undoing' may be needed, and this may at first feel strange, exposing or scary. Unfortunately, what feels physically right is not on its own a reliable indicator of vocal freedom. I've seen pupils surprised and relieved that from their first lesson they no longer felt the discomfort that hitherto they had regularly felt in their throats when vocalising. This wasn't the result of waving a magic wand causing them suddenly to 'get it right', but because no beneficial treatment of the voice causes discomfort in the throat. Progress always implies change. Releasing and developing a voice doesn't mean exchanging one way of singing for another but realising natural potential. Inner unrest may occur because you are not familiar with all there is inside. As you become aware of and familiar with more of yourself, you'll feel and perceive yourself differently. If the liberating process is to be effective, it requires a whole-hearted, whole-minded and whole-bodied commitment.

Equipment

If one department in training fails to pull its weight, the whole will founder. The list below offers some guidelines for committed and effective work.

1. This is your individual journey, so there's no point making comparisons with anyone else. You have your own capacity and special qualities. Search for these within yourself and enjoy them. You begin always from where you are.
2. Observe and try to understand the nature of the singing voice, and thereby its logical and liberating process. This will ensure that you avoid major pitfalls and unfulfilling end-gaining. Learning to sing is in part an act of faith, but you must differentiate clearly between your teacher's role and your own, assuming responsibility accordingly.
3. Learning to sing is a discipline. To make genuine, well-founded progress, you will need to work hard, regularly and methodically like a dedicated athlete or dancer. Consider what is required and commit to it – without this commitment the whole exercise can be a waste of resources.
4. Get fit for this demanding work. This is your responsibility, and your teacher can only be expected to guide your training work in so far as your body and mind are able and prepared to take it. Sleep well, eat healthily, exercise your body, and drink plenty of water.
5. Always begin your training session with an open mind and open ears, free from prejudice and expectations. Be ready to start each session afresh.
6. Arrive at your lesson relaxed and alert. Your energy level needs to be at its highest for both training and performing. No matter what your mood, summon up the energy that you would find if, for example, you suddenly had to deal with a fire, or the concentrated energy of an athlete preparing to run or jump. This high energy level is vital for rhythmical movement, clear hearing, emotional participation and for deep and detailed physical work.
7. Be ready to experiment and take risks. Remember that we only learn to do things by doing them.

8. Be fully awake to the training experience. While it's your teacher's job to train you, it is yours to recognise the experience on all levels, physical, mental, emotional, and above all, aural. Your teacher needs your feedback. Your awareness of the process must be interested and detailed. Reclaim the quality of wonder and develop your capacity for focused attention.
9. Study how to approach exercises so that you can make the most of them (see Chapter 13) and be truly proactive.
10. Child-like qualities are essential for good progress. Genuine curiosity makes you positively want to 'see what's inside'. It can help you overcome fear, and the pressure of needing to get it right.
11. Your voice develops laterally as well as linearly, so all its physical components need to be attended to if it is to grow in balanced strength. Genuine progress is incremental. By appreciating small improvements you can build confidence on a firm base. Having a sense of delay and enlargement can prevent you overlooking crucial potentialising components and help you to train thoroughly.
12. Beware of the illusion of progress arising from the rapid achievement of particular premeditated goals. These are very often gained at the expense of other elements equally important in establishing a firm foundation and achieving wholeness. Beware too of sounding overly mature. A well-balanced voice always sounds youthful. It is easy to make the middle range too heavy, at the expense of the extremes.

Technique

This is a word or concept of which we would be wise to be wary. My colleague Lorna Marshall gave me a nice image of the difference between a process based on reality, and one that's not. In unlocking a door there should be a key as well as a keyhole. Techniques want you to go through the keyhole, not open the door!

We only have to *learn* to sing when it comes to music; then we must draw on the musicality released with our voice combined with learned musicianship. It's only when the parts of a voice are not 'gelling' that we have grounds for believing that this is not true. Since non-gelling voices are the norm, the belief that we must learn to sing is also the norm. When locked up, our singing voice expresses its needs in terms of music, expression and interpretation, seemingly asking to be manipulated to these ends so as to end frustration. Here we have a false solution to correctly perceived needs. While any other musical instrument *is played*, the singing voice *sings,* but does so only if it's whole. Then, given the desire, something to express and appropriate intentions, it sings readily and skilfully. Examples of unphysiological or divisive technical manoeuvres are: taking more than a normal amount of breath, which impedes rhythmical and emotional flow, unwieldy diaphragmatic runs, and aspirated runs which are uneconomical in terms of breath, and decidedly non *legato*.

Techniques at best provide a false, enslaving security. It is the ability to let go and constantly change that brings genuine stability. Genuine security is achieved not in conformity but in gaining self-understanding. Opening up entails imagining oneself in new ways and committing to a journey of rediscovery.

Scale and skills

In physical terms two very small vocal folds are responsible for all our vocal skills and 'prowess' in singing. In the length of these cords (approximately the width of a white piano key) you have two octaves or more. When free, these cords have the capacity to *crescendo* or *diminuendo*, to articulate notes rapidly, and to make a wide range of tonal colours according to emotions and imagination, all simply by tiny adjustments in response to our intentions.

Voices usually open up and develop by degrees, so with patience notes that are at first perceived as 'high' will gradually level out as the elasticity of your larynx develops and its full natural pitch potential (mean tessitura) is realised. Only the *pitch* goes upwards or downwards, while the larynx responds – horizontally – to the notes you 'hear'. Your body's business is simply to sustain the tone with its inherent elasticity, not to 'give it a leg up', or itself adjust to pitch.

Similarly, *forte* and *piano* are degrees and 'qualities' of tension in the same small mechanism: thicker vocal fold tension for *forte* and thinner tension for *piano*. Ranging between *forte* and *piano* is an almost negligible operation when the laryngeal muscles are well integrated. If a louder sound depends merely on a thicker vibration of the vocal folds (more chest register) it stands to reason that greater overall strength rather than extra force will be required to reach the higher pitches.

Perhaps it's misleading that it's the base of the folds, the thickest part, which is best able to offer resistance to force of breath. This may explain why it's so often the first part of the mechanism to be deliberately developed. Some techniques are almost exclusively devoted to 'singing in the mask', the 'placement' for chest register. Unfortunately, though effective results (in terms of tonal concentration) can be quick to begin with, voices pursuing this path can become overblown, or take on a throaty thickness or 'snarl', or undue 'weight', limiting both quality and flexibility. Vocal folds' tension must be matched by their stretching which allows for adjustment and proportion both horizontally (pitch) and vertically (dynamics). Force impedes natural vibration. Interestingly, 'bigger' voices don't necessarily 'carry' better. The degree of carrying power for all voices depends on how firmly and finely honed the edge mechanism (the ultimate tone-concentrating agent) becomes.

Learning contemporary music

Contemporary music is often written not for the singing voice but for the noises it can make. For a singer singing lyrically-written music, knowing how his voice 'does it' is of no more use than knowing how hands work is for the majority of people. But contemporary music can be dangerously fragmenting or distorting. To minimise possible damage a singer needs to know how to produce non-singing sounds. Any singer who wishes to sing contemporary music that requires anything but a lyrical line should heed the following advice.

1. Approach it from a base of natural vocal balance and strength. Contemporary music can be vocally fragmenting and distorting; if a voice is not healthy such music can quickly multiply its problems. Ironically, the singing voice's ability to make 'sound effects' has often been its undoing!

2. A fully liberated, healthy voice is essential if you want to sing contemporary music and survive. Many voices struggle and fall apart because of the unbalancing demands of 'unsingable' music. The saddest consequence is the increasing split between the person and her voice, leading to the loss of her unique tonal quality and the ability to express herself directly and fully in sound.

3. The singing voice needs a varied musical diet, which generally means singing a variety of styles within the classical genre 'spiced up' with lighter fare, without distortion or inhibition. Singing voices are capable of much variation, and to develop must stretch themselves in terms of colour, range, weight, agility, character and expression. If you sing much contemporary music you need to frequently return to music that reminds the voice of its inherent lyrical qualities.

4. Learn pieces of contemporary music, particularly the rhythm, tempi and timing, before attempting to vocalise. Also memorise short sections (even two or three notes only at a time) so that you can start vocal work in a spontaneous manner.

5. Always keep as true to the singing voice as you can (as you should for healthy speaking), and learn unfamiliar effects from someone who knows how to produce them safely. Some contemporary scores have effects that encourage 'ballistic stretching', sudden sharp, jolting movements which tend to be forceful and can damage muscles. Rasping throaty sounds and prolonged whispering should be avoided at all costs!

In between lessons

The following are some of the many other things to attend to if you want to sing well and fully enjoy the process.

Fitness

Our body is our voice, so what we do to it or with it can make a significant difference to our singing, short and long term. General physical fitness is prerequisite for successful training, but the singing instrument needs to be flexible throughout its structure – beware of body building and getting muscle-bound. The breathing system (throat and thorax) reaches to the pelvic floor, but work on the abdominal muscles (as for example in sit-ups) can be an impediment to specific vocal training.

Speaking

Outside the studio you probably do more talking than singing! Bad habits gained in the course of learning to speak contribute to the breakdown of the singing voice, and are a major reason for needing singing lessons. The vocalisation we practice when we're not singing continues to have a significant bearing on the healthy functioning of our singing voice. A singing teacher can help you to speak better, but you must practise the new pitch, sharpness, animation or whatever is required whenever you speak. This might sound strange to begin with, especially as speaking has become a major part of your identity, but get friends and family to support you. Your singing will have a much better chance of improving if you continually practise speaking healthily.

Aesthetic sense and musicianship

To become good musicians we must become acquainted with all kinds of music and observe other performers. Singing is a creative act on the musical level as much as on any other, and we need to feed our imaginations and cultivate a feel for movement, shape, form, proportion, structure, colour and quality, timing and so on. Much can be learnt by observing other performing arts: an actor's dramatic timing, line-shaping or use of space, or a dancer's grace and sense of rhythmical pulse. Playing an instrument, especially together with fellow musicians, can be invaluable in becoming a good, sensitive and practical musician. A singer cannot become an artist in isolation or self-indulgently.

Audiences want to enjoy the music and text as well as the way you look and sound. Read up the background to the music you expect or hope to sing, get into the drama (of songs as well as operas), diligently explore the subtext, and become acquainted with all the characters, their lives and their relationships. You may be surprised how enjoyable and enlivening this can be and you will certainly become a much better singer for this work.

The rhythm of life

By rhythm I don't mean 'beat'. Rhythm isn't metronomic but elastic and fluid. There's no better way of appreciating this than by dancing, and awakening the rhythm inside by listening to your body. This can reconnect us to the pulse of life upon which all truly communicative physical performance is based. Above all, rhythm is deep within us, even at the cellular level. In beginning a piece of music we connect ourselves to that piece's rhythm by embodying the rhythm of the introduction or anticipating the pulse before singing, rather than jeopardising our freedom with a measured breath. In other words, we are already 'dancing' or pulsing before the music begins.

Dancing helps us to understand and make use of our relationship with gravity: can you feel light, graceful and playful without levitating, or feel grounded without feeling burdened or dragging your feet? In *A Soprano on Her Head* Eloise Ristad talks about the connection between crawling (an important phase in our physiological and neural development) and our sense of rhythm. If your sense of rhythm isn't what it should be, check out how you crawl. If it's with right hand and right knee together, and then with left hand and left knee together, you need to practise cross-crawling (right hand with left knee and left hand with right knee) until this is well coordinated, with the respective hands and knees contacting the floor at precisely the same moment. 'Be patient' writes Ristad, 'for it may take more time than you realize. When it feels comfortable going forward, use the same pattern to crawl backwards, and then sideways.' [16.3] This can significantly sharpen and naturalise your sense of rhythm.

Hearing

> *The beginner's mind is an open mind, an empty mind, a ready mind,*
> *and if we really listen with a beginner's mind, we might really begin to hear.*
> *For if we listen with a silent mind*
> *As free as possible from the clamour of preconceived ideas,*
> *A possibility will be created for the truth . . . to pierce us.*
>
> Zen master Suzuki-roshi[16.4]

The singer's ear is vital in the process of effective vocal work. This can be a difficult and muddled concept: the way the ear naturally monitors the voice is not conscious, but we cannot expect our ear to monitor something with which it has lost contact. We have to find a way back to a state in which the ear again recognises our singing voice. What should the singer go by in terms of aural monitoring while this is happening?

Vocal exercises designed to restore your voice to its singing condition are not only searching for your sound. They are 'sounding' your ear, probing it for the knowledge it was born with. While physical restoration is going on, aural connections are automatically being re-established. Meanwhile, of course, the singer experiences other more conscious sensations:

1. **External hearing.** Singers are often surprised when they hear their recorded voice, which only goes to show that, liked or disliked, this was not how they heard themselves while they were singing. Singers must trust their teacher's constructive, affirming feedback so that aural-vocal connections are re-established with minimal external distraction.

2. **Numerous physical sensations**. These should only be entertained in conjunction with sound that is clearly perceived – otherwise they should be ignored unless there is undue discomfort.

3. **Stirred emotions**. These should be noted and permitted. Emotion should also be used judiciously as a voice freeing-connecting device. Emotion is accessed by the release of muscles (including, crucially, the diaphragm) which are preventing feelings from expressing themselves. This prevention is a safety mechanism, often so well practised or firmly in place that we're not even aware of it until our grip loosens. Movement is effectively committing to emotional expression. Your teacher is not out to make you cry, get angry, or feel embarrassed, however. The release is necessary if your voice is to be free, and so is identifying with the resulting sound if you are to fully own your voice and sing with authority.

4. **Value judgements and thoughts about 'how?'** These are often the first on the scene (analysing, interfering, coercing, controlling) and are difficult to shake off. They're accompanied by preconceived notions about how the voice should sound and feel. These notions and judgements close the mind to new experiences and the body to new sensations by censoring and censuring any new emotions that dare to appear or unfamiliar sounds that slip out.

Listening to other singers

We hear a vast quantity of vocalisation which is far from healthy, absorb these sounds subconsciously and are almost bound to imitate them to some extent. It's important to listen to the best examples of vocalisation that we can find. Nowadays voices are often stereotyped, with very little individual quality – the unfortunate side of having access to so much recorded music. There is no updated model of the singing voice that we're all born with, so it's interesting to look for examples of singing from the time of the earliest recordings, making comparisons (for example) between singers of the first 50 or 60 years of the last century and since. Many young singers have heard only of Caruso and Callas. This is a shame, when there were so many fine vocalists in between. If Renata Tebaldi was said to be Callas's rival, for example, it's easy and instructive to hear why. You might try listening to the Swedish tenor Jussi Björling, the baritone Riccardo Stracciari, the American soprano (of Italian extraction) Rosa Ponselle or the mezzo Ebe Stignani, all singers with great vocal facility, and their own special individual quality.

Listening to good examples of vocalisation should be our constant aural nourishment. *Bel canto* singing reached its zenith before recording was invented, but occasionally more recent singers (such as Montserrat Caballé and Mirella Freni) have achieved a degree of excellence which gives us an idea of what *bel canto* might have meant. The signs to listen for are ease of delivery, individual quality and evenness of tone throughout the range, great flexibility in movement and dynamics, a flawless *legato*, integrated emotional expression, meaning in the sound itself, and clarity of diction. These are all interdependent facets of a well-integrated whole.

Other obstacles to liberating the singing voice

Visual elements

The visual elements of vocal music (notes, bar-lines, dynamic markings, text and so on) appear to be instructions that we must obey to the letter. Accordingly, the intellectual process nourishes our need to analyse or otherwise unhelpfully control what's going on vocally while we sing. The freer our voice is to respond to expressive intentions and the sooner we can release ourselves from notational constraints, the greater spontaneity we can achieve and the more fluid, flexible and meaningful our performance can be.

Obstacles to sound progress often reinforce one another. For example, pursuing thought-directed elements of performance as described above while at the same time trying to liberate the instrument that they depend on often leads to thought-directed vocal techniques, which are equally constraining. Unphysiological controls are often linked to a lack of trust in ourselves or in the training process.

Excuses

Singers, like human beings in general, often have many excuses for a lack of success. As Robert Fritz puts it 'Some people seem to enjoy concocting a dramatic explanation for why they can't have what they want. The love of reasons for why you "can't"can supersede the pursuit of what you want'.[16.5] This can be a genuine feeling of inability, an equally genuine feeling of not deserving or some other irrational reticence. Related to this is the pursuit of perfection, perhaps so as to be accepted by others. This can produce a vicious circle: a singer sets herself up for disappointment because she's bound to fail to be perfect. She can't accept that the way she is or what she does is ever good enough so she tries even harder to succeed, often in ways that are not true to herself or her voice. In this way she blocks herself further, thus proving to herself that she's not good enough.

I mustn't!

Our inner censoring voice can not only make us feel ashamed of or guilty about our pain or imperfection, but can also cast doubt on the acceptability of our personal power and joy. Inasmuch as we don't reveal ourselves in connecting with others, we isolate ourselves. What or how much of ourselves must we reveal in order to be heard or accepted? In liberating our singing voice, we overrule our inhibiting censor since we are not only obliged to open up to ourselves and embrace who we are, but are encouraged to share this with others.

Fear

Singers sometimes fear success as much as failure. While fear, which is by and large irrational, can galvanise us into action, it can also sap our energy and reduce our intellectual capacity through its effect on the autonomous and central nervous systems. Whether psychological or educational, whether well-founded or not, fear is a fact of life. A positive way of dealing with it might be to see it as a friend that warns you, not that you cannot do something, but to what extent you need to dedicate yourself to achieving it. If you really want to do something, however afraid you may be, you must give it 100% of your attention and skill. The measure of difficulty is related to the dedication and application required. If you're prepared only to give 80% of your dedication, your fear tells you that you should not expect to succeed. Seen in this way, fear can be a 'diamond' rather than a 'demon'. Resistance to healthy change, however, can be visceral, trapping energy and literally incapacitating the body. It is pointless to deny fear. We have a better chance of using this energy to our advantage if we listen to it and accept it rather than resisting it.

Ideals

Our ideals should not obscure practical sense. The ideas of 'getting it right' or the 'quick fix' are both unrealistic. It may be tempting or seem easier to relinquish responsibility to an authority or 'accepted way' that is persuasive, but the process of liberating a voice is not one of solutions or prescriptions. Setting out to 'produce performances' will not lead to a satisfactory outcome. Expectations can be blocking, setting us up for disillusionment. Stubbornness and adherence to preconceived notions block new ideas or ways of working, and delay integration generally. Impatience causes corner-cutting and an unwillingness to explore. Indifference can be adopted as a defence mechanism to cope with apprehension or lack of confidence. Lack of self-esteem is an insidious obstacle to progress, while egocentricity (often masquerading as confidence) can blind a person to what she really needs in order to 'give birth' to her voice.

Physicality

Another obstacle to progress can be difficulties with physicality. Singers are often out of touch with their bodies or negative about them, but singing is a highly physical activity, and expressive as much of our physical, sexual and emotional nature as of our intellect. The body is an astonishing phenomenon. We need to value it, get to know it, take good care of it and enjoy it. Its condition directly affects our ability to both train and express ourselves in performance. Sometimes we are reticent about physical work because of a history of vocal abuse (perhaps resulting, as it often does, in nodules), or a physical accident such as whiplash. Bodies and their voices are fairly resilient, but where there is doubt specialist advice should be sought so that work can proceed without undue worry.

Posture revisited – concerning ease and effort

In *The Potent Self* Moshe Feldenkrais writes: 'In good action, the sensation of effort is absent no matter what the actual expenditure of energy is. Much of our action is so poor that this assertion sounds utterly preposterous.' He writes later that one only has to observe people who have:

> learned to perform correctly mental or bodily actions (…) in order to convince oneself that the sensation of effort is the subjective feeling of wasted movement. All inefficient action is accompanied by this sensation; it is a sign of incompetence. When carefully analysed, it is always possible to show convincingly that the sensation of effort is due to other actions being enacted besides the one intended.[16.6] (see Chapter 4)

Thorough vocal training, followed up by a pupil's everyday self-instruction, can sort out most cases of normal postural misalignment. Even slight adjustments can make a significant difference. Whenever I've adjusted a pupil's head (forward and up) there has been an immediate significant release of the voice. The adjustment in itself is only a start, however. It's not that the larynx has been released so much as that the suspensory mechanism has been realigned, and can now benefit from training. In particular, what supports the larynx backwards and downwards at the nape of the neck has been made accessible. So long as this vital elastic anchoring is released or realigned, its role as an extrinsic laryngeal support can be strengthened. It is up to the pupil to constantly check on this neck-freedom, with due reference to the alignment of the whole body, thereby guarding against relapse between training sessions. So long as serious misalignments exist in the body, singing is bound to be effortful. Misplaced effort, enthusiastically made, can force the body to go even further out of balance. Lorna Marshall once said to me, 'balance is not a state to which we naturally aspire; it's a natural state which in one way or another most of us have lost.'[16.7]

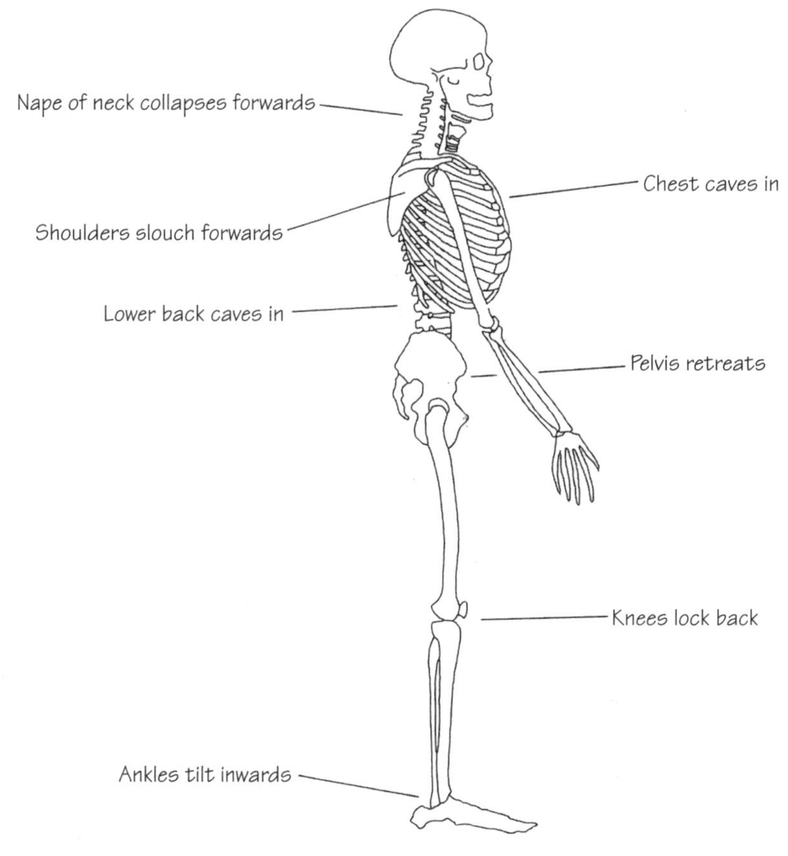

Common postural weak points

You may have observed that attempts to stretch the folds or to open the throat (as in head voice), even with a well-positioned head, can result in a fixing of the throat (a tightness which is usually felt below the jaw at the root of the tongue and at the back of the throat). This is another reason for insisting on imagining and listening for sounds rather than feeling for them. Physical attention might more profitably be focused on lengthening the spine upwards and downwards, thus exploring the postural relation between the head and the pelvis. What *should* happen in the throat and the thorax is facilitated by this lengthening process which enables the head and pelvis to be 'positioned' in a naturally aligned relation with each other without undue effort. In a natural stance we have two lungs full of air. The consequence of efficient breathing *out* is then the full replenishment of this ample supply.

When focusing the tone so that it is sharp and clear, tightness is prevented by the freedom and ability of the throat to 'take the strain'. This is not strain caused by resistance to a bellows-like movement or squeezing of the body, but from the healthy tensing of the vocal folds. A fixed throat is no more a sign of good support than is the sensation that the throat is being supported by the body. A true definition of 'support' might be 'that which renders the larynx free'. The 'support' afforded by the suspensory system works *outwards,* elastically from the larynx, not towards it. The strength is taken onto the body, especially the back, so that the front, from face to toes, is left free, flexible and expressive. We have come full circle to the well-balanced posture which is efficient and free of strain. At this point we can appreciate that there's no special breathing for singing. The larynx is able to work strongly and the breathing system is able to respond appropriately – resulting in a natural reciprocity.

How does your posture measure up in everyday life? It is pointless spending lessons training the singing voice with its dependence on good posture and then neglecting our body in between times. How we conduct ourselves in the everyday actions detailed below is vital to our training progress.

Standing

All healthy 'positions' of the body depend on the centres of gravity within our head (between the ears) and pelvis (at the sacrum). Whenever we're not working with gravity we're putting the body under stress. Since it is not natural to stand still for long periods, we're bound to frequently reposition ourselves. There are many positions that, taken up for a few seconds at a time, may have no adverse effect. It's when we unbalance our bodies in particular ways causing habitual stress or compensation that problems arise. A common example is lower back strain from standing with knees locked back. The diagram overleaf shows typical stances which can stress the body if prolonged.

Sitting

There's only one correct way of sitting, which is on the bones designed for that purpose – the sitting bones. Positioned below the sacrum, these well-padded bones enable us to use gravity in such a way that all strain is taken off the back, shoulders and head. Persistance with this natural way of sitting repays you with a clear head and a feeling of alert well-being. Singing while sitting badly can be vocally counter-productive and unnecessarily tiring. Frederick Husler is said to have once suggested to Herbert von Karajan that he get his players to sit on the front of their chairs (on their sitting bones). The energy and the quality of their sound immediately improved. Most chairs should carry a health warning because they encourage, if not oblige, sitters to sit badly, threatening femurs and tails with pressure that they're not made to take, and causing spines to contort or become compressed, as well as causing poor circulation.

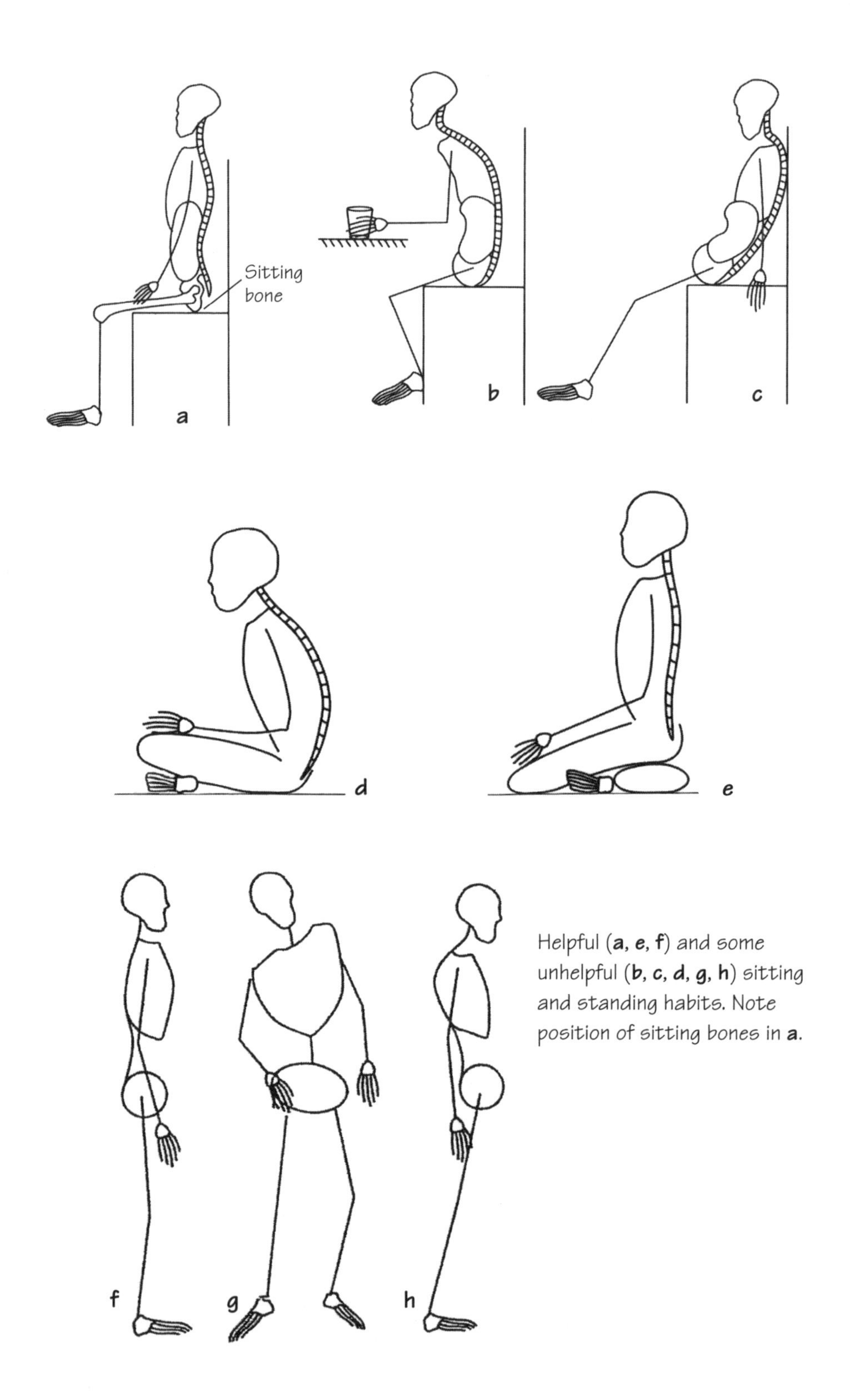

Helpful (**a**, **e**, **f**) and some unhelpful (**b**, **c**, **d**, **g**, **h**) sitting and standing habits. Note position of sitting bones in **a**.

Sitting to standing and back again, we need to lean forward, keeping our back and head aligned, and, with feet firmly planted on the floor, use our legs as levers. This way, no effort is put on the torso or neck. Remember that the centre of gravity is at the *sacrum* which, providing the head is not pulled back, 'leads the way' in these manoeuvres.

Lying down

We spend so much of our life in this position that it is wise to ensure that we do it healthily. Mattresses are often no better designed than chairs. A suitable mattress is pliable enough to shape to the curves of your body and firm enough to support it. When lying on your side, your head should maintain its 'position' in line with your spine, aided by a suitably firm pillow. Lying in the semi-supine position (knees raised, feet flat on the ground) can relieve the body of everyday postural habits that compress the spine, by lengthening and broadening it. Standing up from being horizontal, take care to preserve the relaxation that lying down has encouraged. Without raising your head, let it roll to the side at which you are going to get up. Next let your body follow your head by rolling onto your side. Then use your arms and legs to get you into the crawl position. Sit back on your heels, and then straighten to an upright kneeling position. Finally use a leg to lever yourself up into standing.

Walking

Perhaps this tells us most about our posture. When did you last observe how you walk? Does one hip or shoulder, for example, thrust more willingly forward than the other? Do you make swimming motions with your arms, or work your body as though battling against a gale? Efficient walking utilises the toes as well as the heels. Fully employing feet, ankles, knees and hips, the rest of you can enjoy the ride. Your back is lengthened up and down, your breathing is released (you don't get easily puffed) and your head is free to look around. As in a musical phrase your mind should be 'ahead of the field' so that your motion doesn't drag. Climbing, with head and sacrum aligned, your legs should do the work so that breathing remains easy. Shoes are crucial to how we stand and walk. Constantly bent ankles or squashed feet adversely affect the whole skeletal alignment.

Carrying and lifting

In carrying and lifting, weight should be evenly spread, and in both these activities our legs and arms should relieve our backs of strain. We should beware of physical laziness encouraged by so-called modern comforts. It renders our bodies inefficient and makes us apply force where there is no need. Serious difficulties can arise even from lack of sufficient care in relatively mundane activities.

Assessing the teacher

As a pupil you must understand that your teacher needs you as much as you need him or her. This isn't a question of reputation or money, but simply that a teacher's efficacy is only as good as the pupil's response. You might profit by continually asking yourself 'Am I responding as well as I could, taking the initiative and contributing positively to the process?'

What you accept or reject of a teacher may well depend on your expectations (in terms, for example, of results and comfort) and how far you are prepared to go on this journey of discovery. Expectations and irrelevant agendas can delay progress. Your preparedness to travel depends on your curiosity, courage and relationship with change. These qualities can develop over time. Meanwhile, it can be difficult to discern what good guidance is. Before any such judgement is made, it is crucial to ask yourself if you're prepared (mind, body and heart) to contribute and participate in each lesson, and willing to be guided.

Teachers have their own styles, based on their individual experiences, knowledge and personalities. Recognising this, you can answer the following questions guided by observations, gut feelings and inner searches.

1. **Am I being taken care of?** Has a connection been made between my singing and who I am? Your singing teacher is not your psychotherapist, confessor, nanny, or anyone but your vocal guide. Nevertheless, you and your voice should be treated as one and the same: what's good for your voice is good for you, and vice versa.
2. **What seems to be the teacher's purpose?** 'Shopping lists' of skills (*forte*, *piano*, *coloratura*, a top 'C') will not get you very far if you want to sing well. Can you tell if your 'problems' are being tackled at their roots or is the work cosmetic, 'fixing' problems rather than eliminating them, or overly goal-directed? Are you being offered a 'way' (technique) of singing or expressing music, or is it as though these things are part of you, arising from within you organically? Are the various aspects of your voice being treated as though they were inter-related or separate entities? Bear in mind that it can take time to release the instrumental and expressive qualities of your voice. Impatience for a product will help neither you nor your teacher. Nothing fixed can be flexible. If your voice felt free and flexible at sixteen it should feel even more so at twenty-six.
3. **What's behind it all?** It might help to discover if your teacher thinks of singing in broader terms than the means of singing songs and arias accurately, beautifully or skilfully enough. This will have a bearing on the level and thoroughness of your work, and ultimately on the quality of your communicating skills.

4. **Am I being pigeon-holed?** You may want to know 'what you are', but remember that labels are manmade. Trying to fit into the German *Fach* system can seriously curtail your development. You can fulfil your potential only if you are not limited or moulded at the start. Vocal qualities and skills need time to emerge and develop, and aiming to be a certain type of voice can lead to all kinds of abuse and distortion leaving you unsatisfied because you haven't found your true vocal identity. Repertoire must help the present process of development. Your voice will gradually come into its own and you will discover what really suits it as you go. Trying to make a voice fit a preconceived idea is a denial of responsibility, and it is the pupil who loses out.
5. **Am I being controlled?** Teachers who are authoritarian or dogmatic are usually insecure and limited in what they have to offer. This is likely to limit you and do nothing for your own security. You should have the feeling that your teacher is interested in your sense of self-worth, in empowering you, in helping you to discover what you have rather than imposing what he knows. Training is something active, and it is the experience that counts, not mental know-how, however clever-sounding. Save those grey cells for relevant musical and literary study, and for intelligent musical and communicative performance.
6. **Promises.** Beware of promises such as 'I will turn you into the next Maria Callas' or 'I can make you a singer in six months'. I have come across a number of young singers who suffered due to such empty promises, only to be blamed for their eventual disappointment.

Conclusion

Singers sing for a variety of reasons: some simply love the music, some enjoy the 'physical kick', while others are attracted by the opportunity to perform. Generally people sing (with or without a serviceable singing voice) to express their feelings. Music provides the vehicle and the inspiration for expressing these overtly. A professional singer has responsibilities to the music, the audience and to himself. A truthful performance can be achieved only by liberating and becoming fully aware of the instrument upon which it depends. Since you are that instrument, it makes sense to journey in the enquiring spirit of an explorer, calmly determined to discover the truth in your sound and the sound of your truth.

PART III

The Communicating Imperative

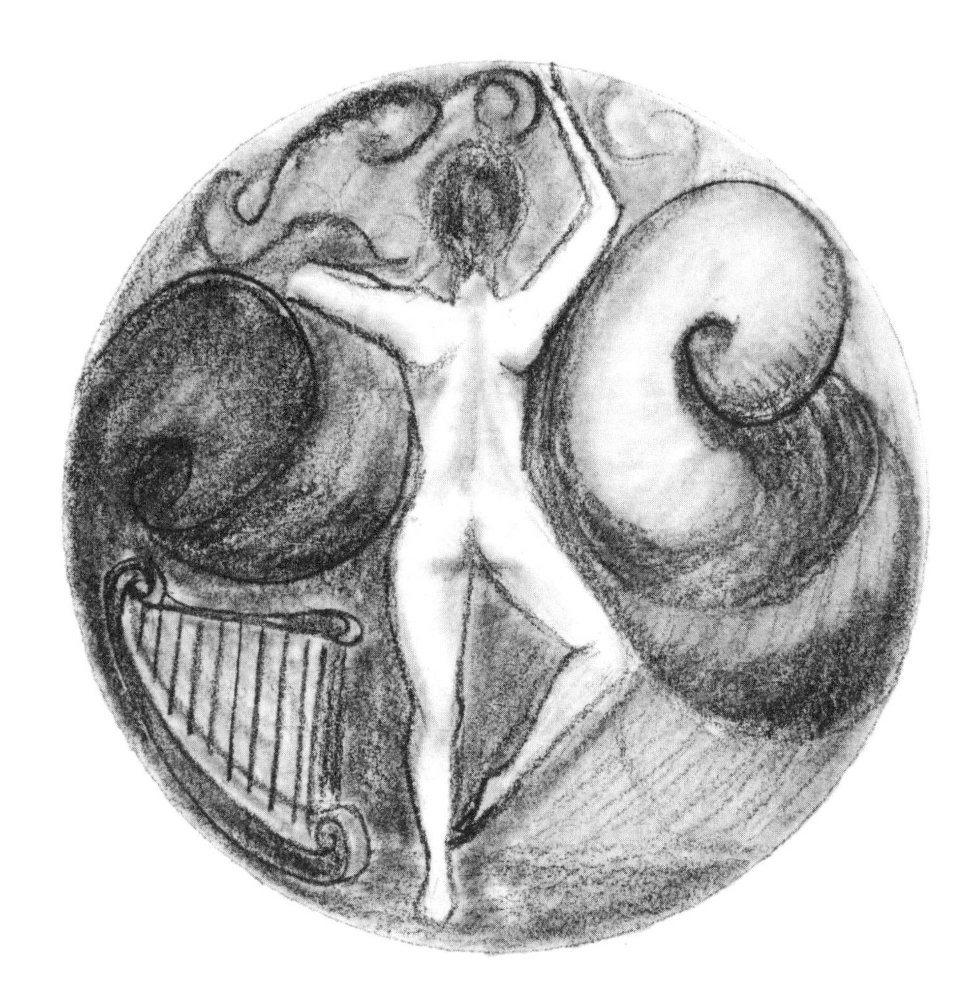

Expression in Movement (Lyricism)

PART III

Chapter 17

Attributes of the Liberated Voice

Introduction

We talk about 'the singing voice' to distinguish it from 'the speaking voice'. While we could define the singing voice as 'that with which we vocalise music with words' there are numerous combinations of words with music and many ways of vocalising. Even in the classical field, singing can mean many different things. We might deduce from this that the world is taking full advantage of the singing voice, and we only have to pick or invent our style. This is a dangerous illusion, based on the fact that the voice we were born with (by my definition 'the singing voice') is extremely versatile. To my way of thinking this is a manifestation of the licence issued, perhaps with the Industrial Revolution, to exploit natural resources without thought for possible negative consequences. I consider the voice trainer's task not so much teaching people to sing in a way that manifests certain skills as regenerating an instrument that has suffered neglect and abuse, the implication being that once restored to its natural shape it can do the job for which it was designed.

I believe the 'desire to sing', as well as our ability to do it, is deep-rooted. Music is a manifestation of our innate if latent lyricism. The singing sound inspires and initiates music. This voice is the one recognised by the great vocal writers, such as Handel, Schubert and Verdi, and the one we seek when we want to sing their music. At the very least, the singing qualities of the human voice comprise one of Nature's greatest gifts to art and enjoyment. We underestimate the singing voice if we fail to acknowledge three of its innate qualities, which are examined below.

1. Its inherent musical skill

We don't have to look far for evidence of the versatility and resilience of the human voice: the 'magical sounds' or 'soul music' of Noh Theatre, imitations of the elements in Inuit 'throat singing', ventriloquism, belting, yodelling, and the *bel canto* of every culture. Because it's normally assumed that vocal skills are learned or imitated we may think that we can pick and choose them at random. However, the extent of the voice's natural skills may not be widely known, let alone that these are parts of an intended whole. Voices are often trained to suit preconceived styles. However, remember that styles of singing often grow out of one person's inimitable vocalising or the imitation of something external to the singer. We cannot say that either the originator or the imitator 'sings well' when he has had to distort or bias his own natural voice in order to suit a man-invented style. Likewise, the imitation of another voice that we consider 'good' or free can never genuinely be our own singing voice. Nowadays, ironically, a lack of natural vocal skill is often the cause of specialisation.

Training the voice holistically is not only fundamental to becoming a well-equipped singer, but decisive when it comes to obtaining genuine results. The 'whole' condition of the singing voice is crucial in freeing the singer's mind, imagination and emotions and in freeing the instrument on which these depend for spontaneous expression. The structure of the whole voice facilitates the complementary musical and expressive skills of which it is comprised. We should encourage the voice to show us what it can do by honouring this holistic structure.

2. What it means in terms of communication

Many singers have intimated that they can only be truly themselves when they sing. For many, learning to sing means 'finding their voice' in the figurative sense. Structurally, as the voice gains its full extent and strength from the top of the head to the base of the pelvis, from soft palate to diaphragm, from spine to sternum, its ability to convey meaning in sound is sharpened and its expressive and communicative abilities are deepened.

Understanding this emotive, penetrating power helps us to perceive links between our individual, universal *voice* and collective human *feeling*, bridging divisions which have been unwittingly cultivated over millennia. The listener responds in accordance with the singer's ability to reach this level of personal-human communication, and with his or her willingness to re-engage with it.

3. Its implications for health

The health of the singer has a direct bearing on the state of her instrument. Logically the healthiest instrument will be the more skilful and strong and the one best equipped to communicate in sound.

The skilful voice

The skills of the singing voice are often obscured by imaginary problems such as the preconceived need for 'technique', a perceived lack of musicianship on the part of a vocally 'blocked' singer, and problems in the combining of words and tone. These arise from a lack of faith in or underestimation of the capability of the voice, impatience, the desire or need to distort a voice for effect, or ignorance.

While any facets of singing music can pose a threat to the integrity and continuity of the sung tone, the major threat is words because their formation is normally so much at odds with the way the singing voice is produced. Before addressing text, let's examine the other facets of singing music as they affect the voice.

1. **Pitch.** Considering the small scale of the vocal folds it is clear that the idea of physically producing pitch is untenable. Separate pitches are achieved by minute adjustments made elastically within the larynx, joined by sliding from one to the next under the guidance of the inner ear. The dexterity that this demands is phenomenal and cannot be achieved by a pair of vocal folds under pressure. To ensure accurate and firm coordination of such minute movements the instrument's structure must be very finely tuned. Intonation is then a reciprocal matter between this precision instrument and a refined musical ear.

2. **Tempo**. A finely tuned voice, with its high degree of economy of movement, can move rapidly and sustain tone with ease. While training, faster tempi are more helpful to progress than slower tempi. This accords with muscle training and sustained effort in general. It is unphysiological to train with slow movements in order eventually to be able to go faster. At slow speeds, muscles tend to fix or cramp. Singers often exacerbate their voice's inability to sustain a slow tempo by filling themselves with breath or fixing their throats. A voice must be free-moving and flexible whatever the tempo.

3. **Rhythm.** An unwieldy instrument will give the impression that rhythm in singing needs 'engineering'. A 'fixed' throat will encourage unwanted diaphragmatic movements which are too heavy for musical precision. The unfettered voice is rhythmically elastic.

4. **Dynamics.** *Piano* and *forte* are merely degrees on a scale of volume, therefore nothing fundamentally different is required to produce either. This was understood by the *bel canto* school when it coined the phrase '*messa di voce*'. Embodied in this phrase is not only the elimination of registers but the voice's natural ability to *crescendo* and *diminuendo* with minimum work. The *intention* of getting louder or quieter is sufficient because the vocal folds and their margins are constructed and interrelated so that this thicker-thinner or thinner-thicker progression can be minutely graded without conscious physical interference, providing tonal intensity is maintained to the end. 'Making' loudness or softness usually results in less clarity and flexibility in both cases, the former through over-use of the thicker part of the folds, and the latter through a reduction in firmness. As with the relative difficulty of

slow tempi, quiet singing is considerably more difficult than loud singing, depending as it does on the hard-won edge mechanism without which the tone cannot be fully concentrated. This specialised mechanism must remain present with its potential for quieter singing when we sing loudly. Likewise, in truly integrated quiet singing the potential to sing louder is ever present. Dynamics are not absolutes, they're expressive devices, and it's important to accept that each singer's dynamic range is individual. Developed over time, this will remain always in accord with the singer's 'size' of voice. The idea that bigger is better or necessary is a reflection of our consumer age. For many singers and for singing in general it has been one of the most insidious trends of the past century.

5. **Colour.** The sung tone is constructed so that it can fluidly bias itself for any musical or expressive requirement in simultaneous response to our intentions. Leaning towards the extremes of the structure – the finest stretching on the one hand, the thickest 'bunching' on the other – in terms of tonal colour, we have an extremely wide spectrum. Add degrees of emotional intensity and dynamics and we have every conceivable shade as well. It's a mistake, however, to think that (except for the odd effect) we deliberately 'paint' our voice while we sing. Vocal colouring is a response to imagination and feelings, and depends on our understanding and appreciation of words, dramas, relationships, subtexts, melody and tonality, and our emotional and intelligent reactions to these things. Suitable colouring is often too subtle and fleeting to be deliberately or cosmetically added. Conscious tonal colouring usually results in vocal distortion, and can also lead to intonation problems Asking a singer to find a new colour should, therefore, be an appeal to understanding, imagination and feelings.

6. **Emotion.** The expressive variety in our emotional voice is also subject to fine integration. Emotion is not something we add, it's something we feel. Being emotionally expressive without feeling is a contradiction in terms. Over-emoting unbalances the very mechanism which makes it possible to move the listener. It's also, paradoxically, only when the vocal structure is unbalanced that the singer himself is disturbed or overwhelmed by the emotions he's expressing. The voice is constructed so that what you feel (hopefully in accordance with the character or text) is conveyed 'organically'. Attempts to be emotionally expressive often lead to exaggerated verbal articulation, as though demonstrating the emotion didactically. A free voice obviates this 'channelling', leaving the face free to articulate words and itself be expressive.

Sing joyfully!

Singers are sometimes urged to sing with joy. When so many songs and operas deal with sorrowful or fierce emotions, one might wonder what this means. A joyful spirit is a good start when entertaining or communicating with others, but what really matters is the structural integrity of the voice, which incorporates the qualities of, for example, both brightness and darkness. Technically, we could relate this to the elevation and depression of the larynx or to 'head voice' and 'chest voice'. A voice is only a reliable conduit for genuine emotions as long as it maintains its balance when one emotion is highlighted from, or in, the context of others. We must sing our sadness therefore with joy in order to maintain the voice's stuctural and expressive integrity. Paradoxical as it may seem, a voice twisted out of shape in an attempt to convey pain will result at best in an artificial simulation of the emotion, which will more likely make the audience cringe than cry. The way the voice moves and inflects when we sing demonstrates its natural affinity with music. Emotion is 'transcribed' into curves or shapes of sound, with varying intensity, which phrase naturally, and effectively realise or 'sound' meaning. Emotional truth therefore is the natural and simultaneous embodiment of the emotional message in sound.

The tonal precision which facilitates, integrates and clarifies the voice's skills is like a multi-faceted diamond. In full flight it catches the light or colour of our meanings, imaginings and emotions as they twist and turn, flutter and whirl, ebb and flow. To make the most of our singing voice, it is necessary to understand that its various facets and qualities harmonise only as a natural whole – they are mutually inclusive. Vocal liberation therefore means not only freedom from constraints, but accessing the lyrical and communicative qualities which characterise this astonishing instrument. Our dedicated and specialised work of regeneration and strengthening is repaid by the emergence of an extraordinary degree of emotional sensitivity and potency which could never be achieved by willpower or musical acumen alone.

Qualities and characteristics of a liberated voice

The list of qualities attributable and available to the fully liberated singing voice is a long one. All must have an equal chance of arising from the same training process.

1. Precision, subtlety and nuances: Singing of a high standard demands functional refinement, clarity and dexterity in all departments and no one vocal skill can be said to be more important than another. Singers rarely get far along the path of genuine vocal refinement because, for reasons already explored, structural foundations are not systematically or thoroughly laid. Sadly, unless they're given sufficient time, singers rarely get more than a glimpse of what might be possible.

The vocal structure's fine-tuning can be done only in the advanced stages of development. With the knowledge and experience of vocal refinement and economy of means can come a closer personal identity with the subtleties of music, text and drama, and a corresponding desire and ability to make more of them. As I read it, the *messa di voce* of the old Italian School was a recognised key to beautiful singing of the most skilful and expressive kind. This finely sourced or centred sound can be the crowning achievement in thorough meticulous training. The elastic quality of the vocal structure permits all departments to interrelate, combine, separate (to some extent) and be highlighted without fuss or ceremony. Continuity of fine flexible tone facilitates clear 'speech', the fluid variation of timbre, pitch, rhythm and dynamics and the integrity and continuity of all facets of meaning and expression. No longer a question of ingenuity or artifice, these qualities arise naturally, spontaneously and inspirationally from a singer's understanding of and involvement with her material; thus the quest for the true *legato* in which all skills are mutually involving and accountable.

2. Phrasing: 'Learning to phrase' is a contradiction in terms – 'learned' phrasing always sounds contrived and unsatisfying. The singing voice accords with our lyrical nature. While a musical phrase has a structure, its form in sound materialises quite naturally given freedom of emission. Inflection follows suit.

3. Lyricism: This means the disposition to respond to or with movement, shape, form, sound, imagery or poetry, and to what is soulful, elevated or sublime in expression. Lyricism is born of the human desire or need to be expressive of things other than the mundane, the rational, the literal, the worked out, the predictable and the known.

4. Impulse: If singing truly pulses there's no necessity to 'attack' sound or 'mechanise' rhythm. The spontaneous initiation of the sung phrase is always firm and flexible, but can be loud, quiet, delicate, commanding, exclamatory, slight, generous, and of long or short duration according to expressive or musical intentions. Vocal training shows us that whatever the tempo or dynamic, one impulse can suffice for the duration of a normal musical phrase providing its measure is accurately anticipated and the vocal gesture is followed through. The liberated tonal line spins as if of its own volition, carrying both singer and listener along as though inevitably. This is similar to throwing a dart at a target; too weakly and it falls short, too strongly and it crashes. Strength of vocal impulse corresponds to the length of the phrase (distance of target).

5. Projection: Concentrated and freely vibrating, the singing voice travels far and with ease. Attempts to make a voice bigger to this end are misguided and self-defeating. A voice may grow in power through training only in so far as it is fulfilling its natural capacity. As her voice gains its concentrated centre, however, the singer may get the disconcerting impression that it is leaving her or getting smaller. Conversely, the fatter, less well-honed her voice becomes, the bigger it can seem. The singing voice is formed in such a way that it both cuts and fills the space into which it travels. In our singing, energy is generated at or fed to the sound source without force so that the smallest voice projects naturally. This energy is our energy, which incorporates our expressive intention or message. When our voice reaches our audience we are still fully there at its core.

6. Spontaneous meaning and creativity: In combining both centrifugal and centripetal properties, the singing voice is self-organising and creative, simultaneously inward-looking and outward-going. The adoption of vocal freedom allows for the unpredictability and creativity of the unplanned present. This is far more exciting and powerful for an audience than the most perfect reproduction or something intricately planned. Human expression is never the same; if it is it is false. Remember, a sound once made is gone for ever; trusting that it carries the moment's message, we must leave it behind. No expression is more present than the singing voice, which explains why techniques seeking to control the moment can never be entirely satisfactory. At the same time, the singing voice, being a total involvement of the singer, is by definition 'meaningful'. The meaning of a work must be fully understood, incorporated and intended if it is to be spontaneously re-created or re-lived in the moment.

7. Presence: Presence arises from a confident relationship with space and time, movement within and between the two, and the ability to fill them without self-consciousness or resort to force or gimmick. With artifice we can be confident (bold) but not sure (authoritative). Accessing the singing voice grounds you, enabling you to be genuine with confidence, and happy with your own energised presence. A voice charged with present life and full of meaning has astonishing power to connect and to move.

8. Commitment: Vocal freedom makes for deepening commitment. Commitment needs freedom for its satisfaction as much as freedom needs commitment for its attainment. Even in a singer whose voice is not free, commitment is far more convincing than skilful contrivance. Commitment with natural skill is a fascinating and compelling combination of mutually satisfying goals.

9. Instinctive expression: We see in training that the voice responds to playfulness more readily, at least in the early stages, than to seriousness. A free voice can as easily cry as laugh. Add to this the role of the diaphragm, teetering on the brink of a thousand emotions and you have an instrument highly sensitive to the exigencies of a healthy emotional life. Inasmuch as the singing voice is innate, it is instinctively expressive. An audience's response to an unfettered voice is primarily on this 'gut' level.

10. Vulnerability: The singing voice exposes simultaneously the individual personality and the human, a powerful state of honest being with which, although we may fight shy of it, we can all identify. The power of singing lies largely in its quality of exposure.

11. Enjoyment: With a free and refined voice comes the ability to enjoy the whole as well as details, connecting and identifying with the content of our sung material and fully entering into and embodying its expressivity and the power of its message. Techniques and manufactured effects can remove us from this possibility and insult our audience's capacity for enjoyment.

12. Duration: In a well-integrated voice the expenditure of energy and breath is minimal. Once a phrase is concluded, the breath is automatically replenished and, in accord with our intentions, instantly ready for the next phrase, providing it has been appropriately anticipated.

13. Drama: In terms of the *Fach* system, although it is rarely explained in this way, the dramatic voice is the one in which the vocal folds are maximally employed. Outside this physiological definition, voices normally gain in dramatic power as they develop naturally (another reason not to categorise prematurely). All performers have the ability to be dramatic; hence the term 'drama' for any kind of play.

14. Quality: Quality is difficult to define. One thing is certain; we should guard against being seduced by sounds which are not structurally sound. Experience shows that the *falsetto* ingredient is a major determinant of good vocal quality. Individual quality 'rings true' for the personality and soul of each singer. To impose a preferred or stereotypical sound on a singer is a violation of her self and her integrity. Given vocal integrity, a singer's tone of voice will vary with what she is expressing, but will always be recognisably hers. The performer's skill therefore lies in her ability to interpret and characterise (as distinct from reproduce or caricature), an ability we all possess.

15. Aural acumen: By the time a voice has recovered its natural balance, the ear is taking over as though finally recognising its own role. Acting in partnership with our imagination, deftly directing vocal operations and receiving the result as feedback, it keeps the voice continuously on track as a functional whole. Remember that the ear hears much more quickly than we can think, so we do well to let it get on with what it knows best, concentrating our minds on intelligent music making.

16. Vibrato: A natural vibrato is indicative of the true, healthy singing that was the concern of the Italian master teachers. Equating subtlety with dry details or with a 'straight' line is as wrong as equating emotion with a fulsome wobble. A well-centred voice is capable of exquisite detail and nuances while remaining vibrant with life and feeling. A wobble (vibration broken loose) is neither subtle nor expressive, and an unnatural vibrato makes accurate tuning impossible. It's instructive to observe that well-balanced voices blend in ensemble while maintaining their individuality whereas unbalanced voices cause tonal, acoustical and musical conflict.

Vibrato in a voice is natural for two obvious and very good reasons. Firstly, the cords vibrate in order to produce sound, and secondly, the diaphragm, crucial to both the structural dynamics of the voice and its emotional content, is in constant subtle movement, mirrored by the soft palate. Remember too that the suspensory system is not stationary; its stabilising effect is sustained *elastically*. The fine-tuning of these physical dynamics ensures that vibrato remains natural and doesn't become unruly or intrusive.

The natural vibration of the singing voice is instrumental in its projection, significant of life, and facilitates genuine emotion in vocalisation. Lack of vibrato signifies in expressive terms a lack of life, or lack of comfort with life, as well as a lack of physical freedom. The combination of an unnatural vibrato and emotion can sound comical as well as expressively indistinct. Emotion with a natural vibrato, however, can be profoundly and exquisitely touching. Ironically, someone who is either reticent or over-eager to express emotion is more likely to over-vibrate than someone who feels free to be natural about it.

Vibrato becomes unmanageable when a voice loses its well-balanced integrity: when, for example, inappropriately 'heavy' music is tackled, or when a voice is encouraged to sound artificially more mature or 'operatic'. A wobbly voice is one that is not sufficiently or freely grounded and/or invites deliberate 'support' (always a sure sign of lack of freedom). A healthy voice always retains its flexibility, its incisiveness and its youthful bloom.

17. Intoning words: What use is singing words if they are not clearly heard? The inherent elasticity of the singing voice is one of the material and intelligent qualities that provide both its capacity for variation and its stability. The stability of the vocal tone facilitates the articulation of text. The better centred the sound, the easier the verbalisation and the less threatening words are to the sung line. A free singing voice has fulfilled its pre-speech potential and does not require participation from tongue, jaw or lips. Words can be clearly enunciated because their physical formation and the vocal emission are not confused. Music acts as a synthesiser of meaning, feeling and soulfulness. What purpose would it serve to add words to music if the music was merely illustrative or a 'backcloth'? And if we needed to express only our rational intelligence, words would suffice without a singing voice. Clearly they do not! What does the voice add to this combination unless it's the feelings behind the words or the contribution of our soul to our communication? The possession of a singing voice suggests that we have *something else* to express, less rational, for which words alone do not suffice. In singing therefore, body, mind and heart can join forces in an expression of our whole human nature without reserve or conflict of interests.

PART III

Chapter 18

Attraction and Repulsion

Singing is perhaps the most important field to recapture the vital impulses, the 'caring power', not only figuratively but physically, of the voice in its free vibrations. Singing is the most direct way to understanding and sympathy with our neighbour. It is the most direct way to a co-ordinated existence, joining the intangible and the tangible elements of life as they can only be united in vibrations. We all know that a vibrating body will generate vibrations in bodies similarly tuned. This is really the secret of the whole universe, of consonance and dissonance, of attraction and repulsion.

Yehudi Menuhin (on agreeing to become patron of a proposed 'International Studio for Singers', 1998)

The singing instrument is made up of the whole person. If singers are intended to be communicators, they are surely intended to be expressive on all levels of their being. The singing voice itself is not the music, the words or even the emotions. So that these various facets of the performing of vocal music aren't confused, we must be aware of what belongs to each department. The role of the singing voice entire is to satisfy the need for fusion or agreement between apparently disparate expressive elements. It does this by maintaining its integrity, in which state it is able to respond naturally, lyrically and feelingly. Too many audiences and performers are deprived of the integrating quality of singing simply because the voice is often called upon to be something it is not. The various elements of expression are allowed to get in each other's way or to 'compete' and none is able to be itself. The sound (the voice itself) is commandeered by the mind as a musical instrument to be played and as a means of producing emotion, whilst emotion itself gets tangled in the articulation of words, so that both become garbled.

The mind should concern itself with making sense of the text and the musical notes. If it ventures too far into emotional or lyrical territory, the voice cannot respond in a truly singing fashion. In the 'classical' repertoire our voice serves every style from dry *sprechsgesang* to the most heartfelt lyrical outpouring, from *recita-*

tivo secco to the most elaborate coloratura: every colour and nuance of 'exchange' between what we have to say and what we feel about it. A fully-fledged voice can run the whole gamut with ease if it's allowed to. Communication must have content, motivation, intelligibility and intention, and if we are disposed to communicate we must give ourselves credit for being designed to amalgamate these elements unambiguously.

The freedom and willingness to tap into and transmit the deeper dimensions of ourselves directly brings us closer to humanity in general. 'Depth, inwardness, and union begin a bond which melts layers of caution and fear, bravado and longing as we share somatic truth' writes Stanley Keleman.[18.1] Although singing is predominantly an outward gesture, it's not so much a giving as a sharing one, not so much selective as all-embracing. It was once put to me that 'growing up' was neither hiding (modesty) nor showing off (confidence), but the desire, ability and willingness to share. A voice veiled or held back gives little away, while thrust at the audience it repels. By means of the subtle dynamic inter-play between expiratory and inspiratory forces, the well-honed voice flies freely outwards with a generosity of spirit while simultaneously drawing the listener into its personal domain. This seems to create a tension of attention and expectancy, a kind of dialogue. True communication feels personal to the listener, who, instead of being 'sung at' feels included and necessary. The listener's response signifies interaction. 'Our basic humanity', writes Keleman, 'depends upon this feeling of connection'.[18.2]

Non-verbal communication (a look or touch for example) can be potent in the extreme. There's a moving story about Dmitry Shostakovich which illustrates the quality of silent communication. The composer, who was given to fits of depression, had a good friend in Mstislav Rostropovich. One day, depressed, the composer rang the cellist and asked him to come round. Rostropovich went to be with his unhappy friend, and they sat in silence. After several hours, the composer thanked his fellow musician for coming, saying that he now felt better. What was important for Shostakovich in his suffering was the *presence* of his friend, and whatever that meant in terms of care and 'cradling'.[18.3]

A speaking voice which is narrow-sounding or nasal will have difficulty in expressing 'I love you' with any feeling in the sound. At best it will convey something by inflection, but any attempt at passion is likely to result in a whining or aggressive sound. Take the words away and there'll be no message of love. Paradoxically, this speaker might produce a better result whispering, removing the core (the heart) of his voice! But would this be believable? A soft-centred voice, by contrast, has a job expressing hate with any conviction – gestures might endorse the words, but no amount of fist-shaking or eye-flashing will make up for the lack of vocal authority. Greater passion in this voice will probably cause the throat to tighten, restricting the vocal expression even further.

In much singing we can only guess what the singer is trying to express. Some voices sound monotonous or angry whatever they are singing about. Other voices sound so round and warm that you cannot believe the 'darkness' behind their words. Words in singing are also often indistinct themselves. Should the listener then rely on the music, physical expression and sur-titles for comprehension?

The true singing voice is so much the whole person that it is capable of conveying emotional convolutions and complexities, knowledge that yearns to be known, knowingness that needs no comprehension, and meaningfulness which defies explanation. These cannot be satisfactorily communicated in any other manner. Music itself has similar qualities, which is why it is said to be a universal language, but there is nothing as directly and deeply personal as singing.

Projection

As long as we endeavour to physically project our singing (sound or text), we put up barriers to communication by dampening natural vibration. A genuine desire to communicate encourages us to make contact with meaning and feeling deep inside ourselves and draws the listener into the world we're desirous to share. 'Projection' can therefore be defined as 'the genuine desire to share' in action.

If the economist E. F. Schumacher had been a singing teacher he might have coined the phrase 'small is powerful'. [18.4] It is amazing the lengths singers go to produce a bigger sound. However, when it comes to the 'art' of singing, communication (and recording) and the projection of the sound per se, what is important is whether a voice and its message carry. The 'size' of a voice is predetermined by nature, so any attempt to exceed it will by definition be counterproductive. The carrying power of a voice is a matter of the natural vibrations resulting from balance. While we might insist that volume (in terms of loudness) is increased by greater incorporation of vocal fold tension, we should understand three important things. Firstly, tension must always be balanced or 'accomodated'. Secondly, chest register by itself doesn't fully concentrate the tone. If a voice's carrying capacity is to be fully realised there must be room for the edge musculature. Finally, the *falsetto*, which in its role of stretching the mucous membrane produces the 'conditions' for a fully integrated and flexible *forte*, is, through virtue of enhancing harmonics, the very ingredient which is most responsible for quality of tone. 'Big' guarantees neither projection nor beauty.

Sexuality and communication

Early in life, we experience our sexuality as an energising force in our physical and emotional development. Assuming your sexuality, you are more complete. The sexual aspect of the voice, though not necessarily as self-evident as the physical and emotional aspects, is intimately bound up with them. Fully present to itself, 'the sexual' can bring anything alive and make it creatively exciting. If it is too hidden, blocked or contained, life goes dry or withers. What interests me is that sexuality and sensuality (as distinct from general sensuousness) seem to be intrinsic to human communication. As the French philosopher Maurice Merleau-Ponty wrote: 'there is interfusion between sexuality and existence, which means that existence permeates sexuality and vice versa'. [18.5] We may not always be conscious of how we communicate on the physical-sexual plane, but as performers we must understand that our aura and physical conduct speak volumes. We can expect our sexuality, as a fundamental part of our lives, to deepen our performance.

In singing, our sexuality is celebrated in its creative qualities and power of interaction. Whether we embrace it or are uncomfortable with it can be significant in our ability to communicate decisively and convincingly as performers. This isn't only because sexuality is the subject of much poetry and motivates many operatic characters, but because being sexual is not so much a biological expedient as an all-pervasive dynamic of selfhood and relationship. It is something which drives us, part of what attracts us to one another, or repels us, even if this is not overtly or consciously expressed. Laughter which attracts, fear which makes us tremble, aggression which repulses, tenderness which caresses, all contain a sexual component. In singing, our sexuality can communicate or facilitate genuine intimacy, charm, passion, *frisson*, warmth (if not heat), grace, child-like excitement, seduction, and so on. It makes for performances which are captivating, urgent, coquettishly sparkling, wickedly witty, oozing with sap – in short, performances with 'personality'.

On a psychological level, repressed sexuality can undermine or diminish personal authority. This suggests that to be a genuinely effective communicator we should come to terms with our sexuality until it is well-integrated and can truly affirm and empower us. The more we come to terms with this natural potency, the more confidently we can take responsibility for the very personal 'creation' called performance.

Our sexual integration or non-integration is indicated, or symbolised, in the structure of the singing voice. The balanced voice in either sex is a composite of 'feminine' and 'masculine' sounds or characteristics. As in the Jungian concept of individuation, in which the *anima* and *animus* balance and complement one another, so in the voice these characteristics, when fully integrated, bring about 'wholeness'. The sounds are mirrored in the structural dynamics of breathing. Put simply, we have the in-drawing receptive feminine creating, so to speak, an acoustic resonating vessel, complementing the outward-directed, assertive masculine, creating a penetrating core.

Strictly speaking, communication that's merely a cerebral exchange without personal significance doesn't exist in singing. It's important not to deny the basic essentials of our being when we come to perform. Body-emotion-sexuality-rhythm lend each other being and purpose.

Singing often lacks 'seduction' or persuasive sap because the singer won't entertain the idea of public sensuality, but confusion reigns when the voice belies the physical sensuality of the singer. In relative terms, it's easy to look or act 'sexy', but genuine, potent sexuality finds many complex ways to express itself. Since, as George Santayana puts it, the sexual instinct lies 'half-way between vital and social functions',[18.6] we can expect it to take on quite different guises which may on the surface have nothing to do with sex as such, but everything to do with living, creative communication.

The most obvious source of physical sexuality or sensuality is to be found at the level of the pelvis, which houses our procreating organs, is the centre of skeletal activity, and the seat of rhythm. It is here where we feel the primary impulse of the dance, where the crucial junction or divide between what roots us to the ground (the origins of our being) and what elevates us to the sky (what we aspire to) is located. Our sexual being and our rhythmical being are clearly intimate partners in the dance of life. Musical education that discourages physical movement or playfulness stems the flow of vital sensuous energy. In addition, a pelvis which is somehow segregated from our being restricts our ability to breathe. There's no greater inditement of the sexual taboo than this deep diminishing of our personal vocal power.

The listener

The audience plays a defining role in the holistic activity we call singing. Listening is often seen as a passive occupation and perhaps some audiences expect to be 'spoon fed' or merely entertained. My own observations as a listener – as student, teacher, member of audience, socially or in relationships – is that this role can be highly participative, demanding, enlightening and decisive in outcome. A 'good listener' is receptive and unprejudiced. In singing and in listening to singing we have a choice whether we truly or fully participate or not. A singer intent on simply 'telling' his audience something will get a favourable response only from those who want to be told, whereas one who shares what he or she is saying or feeling is likely to engage his audience in the experience.

What is certain is that everyone has his or her own reasons for listening to singing and can respond only to what is presented. These two factors seem to me to be crucial – a singer is singing to individuals, and his responsibility is to the individual rather than the mass which, after all, is made up of individuals. For an audience to be 'comfortable' in the sense of being receptive and prepared to fully engage in the experience, the singer must also 'be at home' with what he is doing, fully in command of his material, without extraneous props or distractions;

anything from a music stand to technical preoccupations, irrelevant gesturing or singing to the gallery.

If I thought (as the conduct of some performers suggests) that all the audience wanted was to be impressed or entertained, I would never have embarked on this quest to understand the nature of the singing voice. I believe that in general people have far more wit, heart, sense of drama, desire for magic and wish to be transported in their souls as well as gripped in their guts, and a far greater capacity for compassion and wonder than they are generally credited with or even care to admit. I also believe that singers have a responsibility to satisfy these human capacities and needs. To fob off an audience with a few loud high notes, or with hundreds at breakneck speed, or with singing which is merely clever or ingratiating, insults their sensibilities and belittles the humanity that we share.

That said, adults *can* be as gullible and easily impressed as children. The theatre director and writer Sir Richard Eyre goes so far as to say that 'we worship physical prowess'. There's no doubt that a full-blown high C can be so thrilling that it can seem to turn an otherwise mediocre performance into a triumph! Eyre writes that, 'it's odd how audiences are so easily contented with performances which, as Hamlet says, "imitate humanity so abominably".' [18.7] Content they may be, especially if they know no different, but perhaps the real problem lies in the *imitating*. Imitation, however good, is unlikely to resonate very deeply with anyone.

People have different capacities for receptivity and enjoyment, responding on different levels to different degrees. However, assuming that they're open to what the singer has to sing about, and that the singer is prepared to share her interpretation with her audience, we can expect a reaction on the level at least of 'Oh!' 'YES!' or 'Aha!' if not tears or laughter, even if the reason is not plain. What true singing does, however, is to offer listeners a 'mirror' in which to glimpse or catch themselves unguarded. The more honest by intention and precise in execution this reflecting glass is, the more the individual listener will recognise in it details and depth of his human self and be 'moved' accordingly. This is surely a measure of the power of the human singing voice.

Emotional involvement

Audiences react to what they are given. An important question for trainers of singers is 'what is the singing voice capable of doing for those listening to it?' The bottom line, in terms of feelings, is that people want the truth, however painful. They want to be able to say, on some level, 'Yes, this is how it feels – that's how it was'. They need to be able to acknowledge and celebrate their feelings, not only satisfy their intellect. This is the difference between 'titillation' and something more profound and compelling.

Emotion in singing is in danger of getting out of control or seeming inappropriate only when it's half-expressed. Fully released, the singing voice provides a safe outlet for feelings otherwise encapsulated in musical and verbal formulae. Thus the verbal, emotional and lyrical combine in a balanced, mutually supporting synthesis which, as long as each element remains intact, does not threaten the singer's own equilibrium.

The power of singing to move an audience *en masse* is dependent (music and words apart) on the genuineness of the singer's emotion. As already noted, the physiology of emotions and that of the voice share the same roots and purpose. This alone tells us that, strictly speaking, emotions in singing cannot be 'faked'. A voice bereft of genuine feeling is a voice without roots, and feeling denied a voice. Furthermore, the singing voice shows us that emotion, words and music can be equally precise and unambiguous; they are not merely facets of a composite art that happen to go well together. In addition to complementing each other in whatever varying proportions they are called upon to assume from moment to moment, shaken together they mix and blend into a 'cocktail' which is the most potent form of artistic expression in sound.

Singing sadness with true feeling doesn't spread sorrow. Nor does sharing the sadness make us look foolish. When successful it strikes a chord in the soul of the receptive listener, arousing empathy. When unsuccessful it can become maudlin, boring or burdensome. Singing happiness, especially when its rhythmic quality is deeply founded, uplifts and strengthens the spirit of the listener. Superficially done, it's likely to seem saccharine or merely decorative. Anger honestly expressed induces sympathy either for the singer (as the character he's portraying) or for the 'victim' of his anger. Badly done it's likely to result in an ugly noise that alienates the listener.

The emotional energy expended in the creation and rehearsal of a character can be overwhelming. Being well trained provides the physical and mental wherewithal to survive. A singer must then prepare the ground for the development or evolution of her character by various stages, thoroughly explored and worked, and by careful pacing in rehearsals. How deep she goes and how much she gives must be a gradual process, based not only on 'know-how' and goodwill but on the build up of elemental facility and strength. Measures must be taken not to force the voice into doing more than it is able at one time, remembering that it's a conduit for mind, fantasy, feelings and musicality, not merely an instrument on which you 'play' these things, or with which you replace them.

PART III

Chapter 19

Being Fully Prepared

Introduction

One of the reasons that singing is such a formidably demanding profession is that you have to be many things at once: flexible vocalist, knowledgeable musician, lover of language (and able to sing in several different languages), actor, communicator, and creative artist. You have to be all these in public and to command! Each skill must be well honed and polished, and, if they're not each going to suffer to some extent, they must be treated as equally important. This book has deliberately concentrated on the singing voice, because it is basic and facilitates the physical expression of everything else. In the sense that singing is a holistic activity, however, all facets must work for the good of each other. This complementariness makes for the most confident and telling performance.

Stage fright

Almost all the singers I know have suffered or regularly suffer from pre-performance 'nerves' to some degree. I have personally experienced stage fright as a singer, teacher, and public speaker. Although people experience stage fright in different ways, certain symptoms frequently occur, including shortness of breath or shallow breathing, holding the body 'fearfully' (including, crucially, legs, diaphragm and throat), a dry throat and mouth, nausea or 'butterflies', sweating, wobbly legs or 'out of control' diaphragm, a quickened heartbeat, mental blanking and faintness.

To alleviate symptoms, singers go through routines or rituals, do relaxing exercises, meditate, 'let off steam', carry lucky charms or pray. Others resort to medications, many of which however have deleterious side-effects.

In anticipation of the unknown, a degree of nervousness, apprehension or anxiety may be considered natural, but their negative effects can be minimised if we tackle the problem from the outset of training. Four factors previously discussed need attention and practice from the moment singers decide to take their singing seriously: fitness, breathing, self-esteem and preparation.

While there are examples of famous unfit singers who give good performances, I have observed that they do not do themselves or their instruments justice. People who are fit generally seem to feel good about themselves and prepared to meet life head-on. A positive outlook goes a long way to alleviating the debilitating effects of 'nerves'.

The better a singer's habitual breathing is, the less prone he is to suffering the effects of anxiety and the less need he has of 'special' breathing exercises to relieve it. Efficient breathing is a reciprocal matter between the throat and the body, each being of equal importance and therefore working as one. Without this 'oneness' the system is more vulnerable to faltering or seizing up. As a singer's breathing develops towards a stable throat-body partnership he gains confidence in the ability of his instrument to 'pull off' a performance come what may. A sense of reliability obviates any tendency to compensate for imagined inability or inadequacy. Deep breathing doesn't have to be called upon specially, since it's 'with us' all the time as a manifestation of our fitness and completeness of voice.

A surprising number of singers seem to suffer from low self-esteem, which, though not always immediately obvious, inhibits their performance on some level and negatively affects the way they judge or value it. Pretended confidence usually betrays itself. Conditioning, as part of our upbringing, can affect our attitude to ourselves, limit self-knowledge, and lead to prolonged suffering of the spirit, which has had its wings clipped. In the Foreword of a simply written but deeply challenging book on awareness by the Jesuit Anthony de Mello S. J., his editor, Francis Stroud S. J., quotes one of de Mello's stories:

> *A man found an eagle's egg and put it in a nest of a barnyard hen. The eaglet hatched with the brood of chicks and grew up with them.*
>
> *All his life the eagle did what the barnyard chicks did, thinking he was a barnyard chicken. He scratched the earth for worms and insects. He clucked and cackled. And he would thrash his wings and fly a few feet into the air.*
>
> *Years passed and the eagle grew very old. One day he saw a magnificent bird above him in the cloudless sky. It glided in graceful majesty among the powerful wind currents, with scarcely a beat of its strong golden wings.*
>
> *The old eagle looked up in awe. 'Who's that?' he asked.*
>
> *'That's the eagle, the king of the birds,' said his neighbour. 'He belongs to the sky. We belong to the earth – we're chickens.' So the eagle lived and died a chicken, for that's what he thought he was.*[19.1]

This story is simply saying that you may unknowingly be a 'golden eagle', unaware of the heights to which you could soar. For some, unaware that they are not fulfilling their potential, the 'conditioned' self may feel safe. When it comes to performing, contradictions may manifest themselves in inhibiting or rebellious ways: talented and with 'something to say' or a deep desire to sing or communicate, a singer can nevertheless deny his ability, undermine his own work, or 'make up for' lack of confidence by 'putting on a show'. No amount of applause will give us self-esteem if it is for the clothes we are wearing rather than for who we really are.

There are many life-affirming and confidence-building therapies. The Feldenkrais technique (awareness through movement), for example, is deeply effective in helping a person to re-connect to his original self (see bibliography). One of the most enjoyable and powerful therapies I have come across is 'clowning', a wonderful way of releasing and renewing valuable creative qualities by re-authenticating 'the child within'.

Effective preparation depends on many interrelated elements. A weakness in one area can adversely affect the others. Part of the problem is failing to realise that successful singing is not a list of desirable aspects from which one can choose, but various components that potentialise each other into a logical and stable 'whole' that can radically reduce the incapacitating effects of fear.

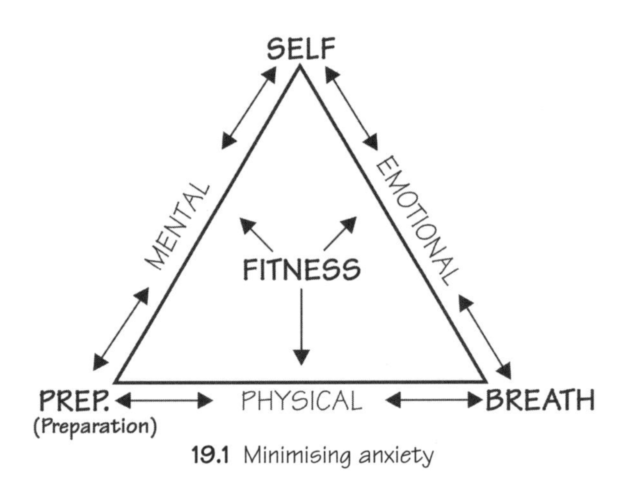

19.1 Minimising anxiety

The unknown

Lack of confidence can arise from fear of the unknown. In a performance your pianist's music might fall to the floor, a mobile phone ring in a crucial pause, or an earring might fall off! One symptom of being adequately prepared is being in command of yourself on stage, having the confidence to be 'normal' in the 'abnormal' performance situation, and being able to deal with practical crises with humour and good grace. On the musical-performing front, singers must be so well prepared that their performance can be real. There will always be some in the audience who are not satisfied, and you will probably always think you could have done better. Being prepared is realising this and still being able to give your best performance.

Performing

In connection with not mixing objectives, I have been asked how I expect singers to become performers and artists while they are not singing. Firstly, I would not expect them to become particularly good performers or artists with faulty, incomplete instruments, which command too much of their attention. Secondly, many performing skills can be developed without singing or even music. From the start of training, a student can, for example, tell stories, recite poetry or monologues, give illustrated talks or even dance, mime or juggle. Such entertainments, which should encourage contact with, or even the participation of an audience, however small, can be instructive, confidence-building and fun. Due attention should be given to presentation and performance, audience response, use of space and so on.

Criticism

The fear of criticism may get in the way of your performance. Most audiences are there to enjoy themselves, not to judge you. Criticism will normally be of your work, not of you personally. Slips aren't necessarily the result of avoidable laziness or laxity, but of unavoidably being human, and hardly merit serious criticism! How 'well' a singer sings is always debatable and critics are often divided – so let them argue! In competitions or auditions you can sing your best and not be accepted by the panel, or sing less than your best and be instantly signed up. What or who adjudicators are looking for is often decided before the event!

Guided in the early stages of his development by people he can trust to give him honest and constructive feedback, a singer must become his own self-knowing and level-headed critic. He must know the difference between constant self-criticism (a symptom of lack of self-confidence) and being objective and fair. While training, a singer can practise being positive by chalking up 'pluses' (however small) in his performance instead of minuses, learning to be realistic about his own expectations according to the stage he has reached, and comparing himself only with his own capacity. A healthy self-appreciation does not preclude healthy modesty. All great singers are highly individual, and the quality of their work continues to be debated long after they're dead!

Solo

In the world of music, let alone outside it, few people understand singers, the nature of their voices and what they have to do to achieve results which are effective and healthy. A singer is very much on his or her own. This underlines the necessity of building self-confidence on firm foundations, becoming fit for the job in every way. A singer must learn to stick up for herself and the conditions she needs in the absence of understanding or in the face of disbelief. This isn't only a matter of musical tempi, an individual's range of dynamics, or the unreasonable (and therefore counterproductive) length of rehearsals, but also of life outside the theatre. For example, it may be difficult for a singer's partner to understand or cope with all the attention that may need to be lavished on the voice, especially prior to a performance, let alone the adulation of fans.

The stress of performing can be immense, and it's not only young and inexperienced singers who suffer. The unhealthy life-style of Enrico Caruso, one of the most famous tenors ever, was probably linked to fear or anxiety. Caruso, who was sometimes physically sick before performances, is said to have smoked like a chimney. Sataloff writes, 'The deleterious effects of tobacco smoke on mucosa are indisputable. It causes erythema (red blotches), mild edema and generalised inflammation throughout the vocal tract'.[19.2] Caruso died of lung cancer at the age of 48, when he should have been at the height of his powers. Fame brings its own stress (measured in media 'hype', and the expectations of crowds and critics), and may bring schedules which are unreasonably heavy even for the fittest. The demands and difficulties of the singing profession tend to catch up with singers as they emerge into the limelight and begin to undergo close scrutiny. Those who are not well prepared for this from the beginning of training are unlikely ever to fulfill their potential or even survive long.

Being fully prepared

The coach and accompanist Jeff Cohen once said of preparation 'Good enough is not good enough'. What constitutes a 'good job' can be roughly summed up as being as clear and truthful as possible in communication, providing an exciting performance, and doing the music and drama full justice. This manifests concern for our audience, for our performance and for the 'work' itself. To be a good communicator, an attractive or compelling performer and a conscientious but imaginative and perceptive interpreter, we need to study and practise the various facets of our art in depth and at length. Too many singers aim straight for a product, or spend too little time on crucial steps or building blocks. Preparing must be a gradual, uncomplicated, planned process, an evolution from something basic or relatively rough through various logical stages of 'construction' to something refined and polished, which enables the artist both to 'be' and to be creative in performance. To be a creative performer you must be master of your material, not subservient to it. The statement 'good enough is not good enough' also implies that performers tend to lose in performance a considerable amount of the good they have gained in preparation (some reckon 20% or more). You have got to be better than good enough to be able to step onto the stage with genuine confidence.

Successful, fully effective performers will explore the background of the work to be performed and study the text long before they begin to vocalise it. The text or poem is after all normally the inspiration and basis of vocal music. The composer expresses words in musical terms with the pace, inflection, emphases and colour he or she feels is apt. The singer must show an interest in this. Having got fully to grips with the text and its meaning and dramatic context (songs are almost invariably mini-dramas) and how these impact on her character, she must endeavour to understand what meaning the music itself lends the words. Texts, their compositional interpretation or transformation, and the singer collaborate in this creative process. The work finally (and only) comes to life in the singer's singing. The singer's task is to interpret a text and communicate it, every single word, and therefore she must discover the composer's intentions. Only then can she bring her own qualities and character and skill to the completed composition, bringing it into actuality. A good performer is not only accurate or 'authentic', but goes beyond the book or schooling, imbuing her work and tastefully enhancing it in terms of text, character and music with her own personality. This renders the piece truly present.

Knowing the words and music well is merely the prerequisite for beginning to rehearse, whether with a pianist, coach or conductor. Such basic preparation can be done at home, thus enabling the rehearsal period to begin productively. Ideas and queries that come up in rehearsal – a level or nuance of meaning, the creation of mood or atmosphere, a re-examination of intentions or subtext, a tricky sequence, questions of style, matters of ensemble, tempi, dynamics and other expressive details – can be noted, taken home and practised or absorbed so that the following rehearsal builds on work already done. Assuming these questions are reasonable and can be acted upon, any self-respecting singer should be able to return to rehearsals ready to implement decisions and changes. Thorough pre-rehearsal work gives you a solid, flexible base from which to do this with ease and confidence, enabling you to enjoy and make the most of the process. Without thorough preparation, rehearsals can founder with uncertainty, conflicts of interest, and tensions, resulting in a mediocre performance of which no one can be proud. With firm ideals shared by all concerned, and given room for negotiation, it's possible to enjoy making music in a confident and truly collaborative manner. Whereas the hard and sometimes laborious work of thorough preparation is amply and appropriately rewarded, cutting corners weakens the work's structure and jeopardises the final outcome.

A significant aspect of the preparation process is the knowledge of where you are in your development as a singer, so that you don't spoil your present ability or lose your identity in the effort to satisfy as yet unreasonable or unrealistic demands, or suffer in the attempt merely to please. Your best performance will be one you can achieve without force, in a state in which you can feel most yourself, and in which your intention is genuine. This is why some untrained voices can be so affecting – the singing is (as far as it goes) genuine and unalloyed. For a trained singer, of whom greater things are expected, self knowledge is of vital importance, even a matter of survival.

Both the process of arriving at a performance and the performance itself should contribute to a singer's overall progress. You cannot add things to your voice, but you can make the most of it without jeopardising its future. I often find myself advising singers to sing 'on their interest' not on their 'capital'. Singers who sing too much or too heavily, or twist themselves into knots in a naïve attempt to produce something better (or to satisfy the unrealistic demands of some conductors, directors or coaches) are at best eating into their capital, and often contradicting the positive hard work they've invested. Such reversal may not be evident before it's too late – I recall the 'tragic' demise of some extremely talented singers of my own generation. Singers must know their weaknesses as well as their strengths and learn to take care of themselves in any situation. There's rarely anyone around who understands what a singer requires in practical terms, hence the non-conducive conditions with which singers almost invariably have to contend.

Spontaneity

It is in the nature of the singing voice to act impulsively, and feelings, strictly speaking, can only be vocalised spontaneously. Potential obstacles to spontaneity are minimised through thorough investigation, understanding and study. If text and music are thoroughly absorbed before any attempt is made to voice them, the best may be got out of a voice at any stage of its development (see Figure 19.2 opposite).

Spontaneity is needed for the immediate, clear onset of the sung phrase, which in essence is the tone stream already in flight, like an aeroplane the moment it lifts off. With the whole phrase appropriately intended from the start, this stream should need no further conscious direction or adjustment before reaching its destination. During performance the voice should be constantly in a state of suspended motion, always alert, energised and expectant. This 'state of readiness' doesn't contradict the idea of vocal spontaneity, because the voice isn't conscious of itself. Unless we think technically (and thereby rule out spontaneity) the free voice is always available. The moment tone is initiated, it is over and gone: the on-going sound remains fresh and is never reproducible. We must give the voice credit for these qualities and not stand in its way. To avoid hesitation we must add to energised readiness clear intention and personal authority.

Continuous spontaneity is best achieved by first getting away from the notes on the page, phrasing in shapes and gestures (each phrase the result of one impulse), and trusting our ears. People's musical ears are generally better than they think. Get a singer to re-read a phrase that he's sung incorrectly and he is likely to remain hesitant in trying to 'steer' his voice around the notes more accurately. Play or sing the phrase and get him to repeat it without looking and without thought, and he usually gets it right, his voice following what his ear hears.

19.2 Basic Logical Song Learning Procedure

***A.* WORDS: The text is the basis of the song/opera extract.**

a) Extract the words from the music and memorise as though for a recitation.
b) Discover the meaning of EVERY word and how to pronounce it.
c) Explore context. Why did the composer use this text? Explore dramatic and historical context. Who is the character? From where has s/he just come? Where is s/he now? Where is s/he going? To whom is s/he singing and why?

Character and drama begin to take shape.

A must be thoroughly mastered before you move onto ***B***

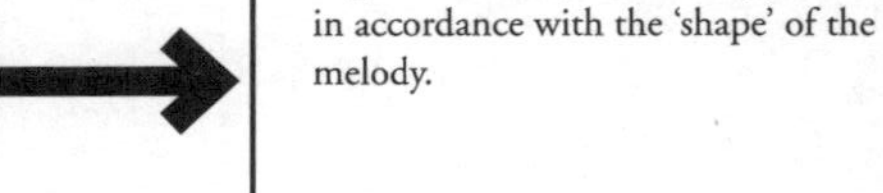

***B.* RHYTHM: This is the framework for both words and melody.**

a) Write the rhythm (only) on a single line.
b) Learn (with indicated or faster tempo, and a good feeling for pulse).
c) Be able to clap or tap from memory – try using one hand for regular pulse and the other for rhythm.

Master ***A*** and ***B*** before you move onto ***C***

C. WORDS AND RHYTHM:

a) Combine these separately from the melody.
b) Pitch and focus your voice healthily.
c) Begin to inflect your voice approximately in accordance with the 'shape' of the melody.

From ***B***, at the same time as ***C***, but not together with it, move onto ***D***

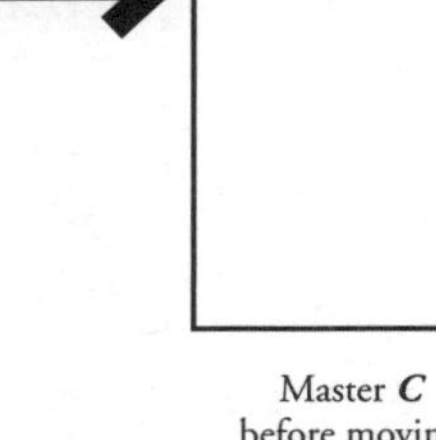

D. MELODY:

a) Learn FIRST IN YOUR HEAD or with use of a piano, without the text, and ignoring dynamics and musical instructions, but taking the melody no slower than the intended tempo.
b) Hum, whistle or vocalise on comfortable vowels, discovering the melody's expressive qualities.
c) Constantly be aware of the pulse.
d) Study the harmony that goes with the melody.

Master ***C*** before moving onto ***E***

E. THE ELEMENTS COMBINED:

a) Now you are ready to combine the elements SPONTANEOUSLY.
b) Precede this (if helpful) by singing only the vowels.
c) When words, rhythm and melody are successfully (spontaneously) combined, you have a solid basis for beginning work on interpretation.

NOTES:

1) Each element must be well mastered before a satisfactory combination and synthesis can be achieved.
2) This procedure establishes a secure basis for spontaneity, avoids a conflict of interests and is an excellent aide-mémoire.
3) With practice and discipline, this logical sequence makes easier and more efficient both your learning and subsequent work with a pianist or conductor. It is a far more effective and calm method than the usual generalised and hasty approach, and makes for:
   a) A greater understanding of the whole
   b) A solid basis for creative progress
   c) A sure basis for confidence and enjoyment
   d) The facility of integrated nuances and detail.

Similarly, counting and measuring can prevent the very thing (the embodiment of pulse) that brings rhythm to life. Vocal gestures must be initiated with conviction, boldly executed, and unhesitatingly followed through like a tennis stroke, as though the sound continues beyond its destination. It is as much the 'following through' as the fluid initiation that brings success. One of the biggest obstacles to spontaneity is inadequate preparation.

Learning and memorisation

Many people have difficulty with memorising, especially if they've not been used to it from early childhood. Singers should practise learning texts from the beginning of training, developing this facility for when it becomes imperative. It will make all the difference when trying to communicate directly with an audience. There's no foolproof way of memorising, but I believe it's not the memorising as such which is the problem but the ability to recall longer-term what has been learned. Our minds are packed full of memories, experiences, impressions, names, languages, facts, ideas which remain filed away, waiting to be accessed or processed with hardly a moment's thought.

Our memory's filing system seems to be fluid, not rigidly compartmentalised, which could account as much for confusion as for clarity, so that one thing can be mixed or associated with another, or even obscured by something else. Mixing, associating or going off at a tangent can happen as if by chance. It seems to me that what makes the crucial difference between easy and effortful access and between clarity and confusion is the associations that accrue in the learning process; it's *how* you learn, as well as in what spirit, that counts. It's not difficult to see that successful recall can depend on the broader process of preparation, practice and rehearsing. Conducted with focus and logic, preparation can lead almost inevitably to clear, easily recallable memories.

A pre-requisite for learning music is taking an interest in it. This may seem obvious but it is not always easy. Even if we don't like a piece, however, it invariably has many facets to explore, from general background to tiniest detail. For this we must be curious and eager. Three elements can act as basic references on this journey:

1. **Comprehension.** If we understand what we're dealing with we not only express it better, with more confidence and authority, but are likely to remember it more effectively. If the ideas, images and sentiments of the text and the purpose of the music are not understood, memorisation resorts to rote. This way of learning is baseless and therefore unreliable, resulting at best in something dry or non-committal. Learning without understanding communicates little if anything of value.

2. **Context and perspective.** Our learning needs a broad picture or la in which to take form, a ground plan in which all the elements involved ... 'see' their own and each other's relevance. This will help to determine what is especially important or else irrelevant. Singers often complicate, add to or confuse work by failing to adopt this discipline.

3. **Components.** In performing music the brain must process both sensory paths and the muscular systems that coordinate for vocalising simultaneously. In addition to the physical workings of the vocal apparatus, words, rhythms, melodies and harmonies are fed into our mind. These in turn conjure up images, sentiments, character and so on. On their own they may be complex or relatively simple, but all have their own intrinsic value. Singing is one of those things that taken as a whole we term far greater than the sum of its parts. Even so, it is the parts that make the singing. Thus we must ensure complementariness through clarity of purpose, not confuse issues or try to gain everything at once. We may have to give priority to the more complex issues but must become aware of how all the components make sense of one another with homogenous ease.

These three mutually reinforcing Cs can help us to balance our work and keep us appropriately focused.

Means

The brain seems to work by the association of ideas or images, linking them together or creating mental pictures (or 'mind maps'). When we see the word 'house', for example, it means nothing but what comes into our minds in the shape of walls, windows, doors, roof, together perhaps with associated images and memories to do with our homes, architecture or the realm of childhood. One thing leads to another – it is by these means that we can map out an association of ideas throughout a song or opera.

We can create a 'story line' setting the drama within the context of what has led up to the point where the action begins and what follows. An aria removed from its opera needs such a context for singer and audience to make sense of it. *Da capo* arias (especially those with endless '*addios*' and no attempt on anyone's part to go), although they may constitute a comment or encapsulate a state of mind, often seem quite static, and challenge the singer to find different ways of expressing repeated words – a different intention, greater or lesser intensity, though never merely doing something 'for the sake of it'. Differences must contribute relevantly to the drama of the moment, leaving the audience clearly aware of what a character thinks or feels, and where he or she is heading physically and psychologically. Ornaments and embellishments should suit an expressive purpose, deepening the sentiment or intensifying the drama. To achieve direction and memorisation, the singer must be sure *why* he's singing the aria and (in the absence of staging, with whatever physical associations that might supply) must construct a psychological journey. It can help to create a physical map too, associating each twist and turn of the drama with different movements and positions in real or imagined space. So much is gained by hearing, feeling and relating to the harmony of a piece that all soloists are strongly advised to work especially hard on this aspect of their art. Being a significant part of the aural landscape, harmony can be another vivid aid to memorisation. The harmony might be what makes a melody both interesting and memorable.

Dido's Lament is a short example of the use of repetition and the effect of harmony on vocal expression. Beautifully conceived and crafted, the haunting repetition of 'remember me' virtually speaks for itself; any 'obvious' change of expression will spoil the aria's tragic simplicity. As so often in the music of the masters, everything is there. Singers often miss Purcell's touch of genius in shifting 'remember me' forward a bar the second time around. Simply savouring this change in tonality can produce extra magic. Dido has a journey to make from taking Belinda's hand (in the preceding recitative) to embracing death. There's little time for this enormous step. Many arias prolong the journey to be made during their course. In all cases, singers must be understanding, imaginative and inventive, to keep themselves and their audience 'on track'.

Priorities

Singers often try to put a whole piece of music together from the outset. This is foolish, making for mental and physical conflict between performance components. Details and refinements can only satisfactorily come into their own organically if they have a sound physical and intelligent base. By being clear about our performance components, we can appreciate how they complement and potentialise one another technically and expressively. The piece, the resultant whole of the components, has already been imagined in the composer's head: what we see on the page is merely a code. To make the most of a composition, we must work out the composer's intentions and thereby understand the piece from the inside.

In understanding how and why a piece is composed as it is, we contribute to our 'mind map'. As we go on, relevant connections or associations can be made between, for example, the words (the descriptions or explanations) and the sentiments that underlie them. More often than not these sentiments are clarified in the music. We should ask what emotions the *music* inspires. Do these emotions correspond to the words, or is the music illustrative of a general mood, or of something else in the scene or drama? The more vivid these connections are the more chance they have of being clearly etched into our memory.

The journey through a piece can be made rationally, emotionally, physically, visually and aurally. What is the musical or dramatic structure, what is the logic behind it? What do the various components and stages of the journey make you feel? Say the words and *visualise* their meaning. How do they feel in your own language? How do text and music manifest themselves physically? Translate your work into gestures, choreograph or dance it. Discover how different saying or singing your text feels standing, sitting or lying down. How do the words and emotions reflect in your face? Suppose your song is set in a garden. Can you visualise it? Can you see yourself in it? Can you feel the breeze or smell the scent of flowers? There are endless ways in which you can stimulate your own imagination and bring the drama of your music (great or small) to life.

A singer must become fully acquainted with the harmony, tempi and pulse of a piece, asking what they signify in terms of mood, character, landscape illustration and so on. These too make the composition more present and alive. In the end you will have a journey mapped out, a story to tell, and a logical sequence of events or physical impressions to re-create.

Details

As you become more aware of your 'surroundings', you'll be able to take in more detail. Details are not mere titivation or cosmetics, and they are not absolutes. Indications such as f, p, < > or *rall.* are there for an expressive purpose, to highlight or bring out the drama, to intensify or enhance the emotional expression. They help to make a piece telling or accessible. It's the performer's job to find out what difference details make. The 'shape' of a phrase can make as much difference to the shape and pace of the drama as harmony can to mood and sentiment.

Many singers take uninvited liberties with music. This is perhaps most obvious in the *verismo* style, where, without due regard for the composer's wishes (let alone good taste!) emotion can take over in an exaggerated or slapdash fashion. Take the endearingly simple 'O mio babbino caro' from Puccini's *Gianni Schicchi*, for example. There are six high A-flats in this short outpouring, but only one – in my Ricordi edition the 'men' of 'tor*men*to' – has a 'natural' emphasis. It makes no sense to pause on any of the others, and an emphasis on 'struggo' only takes away from that of 'tormento'. Puccini has already carefully extended 'Dio', so what is to be gained here by a *rallentando* and pause? The last high A-flat is for the 'pie' of '*pie*tà', a word with its own accent. To slow down here spoils or renders ineffective the subsequent *ritenuto*, reserved as it is (in the accompaniment), for the last six semiquavers of the same phrase, and leading naturally on to the conclusion. A brief glance at the words that go with the other high A-flats makes it plain that it's nonsense to hang around on them. Try it, as many do, and see how much it detracts from the natural flow of the text and music.

While *portamenti* are understood to be part of the *verismo* style of vocal expression, they are for the subtle exposing of real emotion. The phrase 'Mi struggo e mi tormento' in the example above is often swamped in 'scooping' movements which, even with Lauretta's exaggerated words, have nothing to do either with Puccini or with honest emotion. *Portamenti* need taste and a refined instrument to bring them off.

Another area often given insufficient attention in opera is the relative values given to notes and rests in recitatives, which can make all the difference between one interpretation and another. Singers need to ask 'why this note value on this syllable?', 'why emphasise this word not that?', 'what is going on in the character's mind?', and 'what does this silence signify?' Singers often fight shy of silences, when in fact they can be pregnant with meaning or provide suitable time for a change of thought or action. These are definitely landmarks for our mind map. *Note:* it is important to remember that a so-called photographic memory does not serve as a mind map. Even if such a thing really exists, a singer's mind should not be occupied visualising music in performance!

The personal touch

Crucially, anyone's singing depends for its truthfulness on being addressed to somebody or something: a character on stage, someone in the audience, the singer herself or the audience *en masse*. Equally it could be to an imaginary person or an absent character in the drama. Our song might be addressed to a tree or the moon. Such a focus makes our singing more present and convincing. Invariably communication is truer and more effective if it strikes the listener as in some way personal. Focus of address can reinforce our memory quest.

PART III

Chapter 20

Going Deeper

The song of the Italian peasant woman struck our ears all of a sudden. Perhaps more significantly, it ignited something inside us. Whilst busily picking (and sometimes eating) the farm's delicious cherries, we were unprepared for the breaking of that eternal, blue-warm silence. We were not even in an expectant state. On the contrary, we were absorbed in our little worlds. Yet something deep inside us was quickened in that instant; something ancient and yet strangely familiar resonated with that sound. It seemed to be a spontaneous and whole-hearted response to something deep inside that humble human being. It was not so much the sound to which we awoke, as its source.

Adapted from 'Notes' by the author written in the 1970s

The source of the singing voice

When considering what singing in its truest and deepest sense is about, we must consider the essential relation between vocal sound and its source. When it comes to hearing, it's a matter of engagement between the hearing and what or who produces the sound. This, however, is not simply a sound voiced and a sound heard. It's something more profound than action and reaction, unless we use reaction in a scientific sense. This deep engagement and its effect is in fact uncannily like a biochemical reaction, in which cells are equipped with many surface 'receptors' (in our context, listeners), each a single, vibrant molecule awaiting a partner. The partner (the singer's voice), a chemical with its own properties, called a 'ligand', swims about in the extra-cellular liquid (the air), looking for its receptor, in entering which the 'dance' begins. The partners find themselves in tune with one another and the steps of their dance transmit to the cell information that brings about the beginning of profound changes in that cell.

Although the singing voice has the capacity to move an audience *en masse,* it's to each individual that it speaks. Singers need to bear this in mind if they want to bring together and concentrate the true singing qualities of their voices. Before ɪn happen, a singer must be prepared to *recognise himself* and his humanity voice. The deeper we go in training, awareness and acceptance, the greater capacity our voice has to bring about self-recognition and inspire transformation in others.

The process of regenerating the singing voice is one of overcoming those aspects of civilisation which cause us to be something other than our true self. First we must overcome our mind-sets, our mental or intellectual defences against change or reconditioning. Then, being the instrument that sounds us, we must open up our bodies. Our spirit can then be released through its mouthpiece, the singing voice. The receptive listener is struck at the meeting point of these two deeply human sides of his or her being.

This meeting point is the unification of what at first looks like opposite and unrelated poles. The struggle to unite or bridge these divisions in the voice represents our struggle to individuate. On succeeding to reunite these mutually dependent aspects of voice, we find the instrument perfectly attuned to express the essential indivisibility between our individual (undivided) and collective being. This unification is what can make us such powerful communicators. The final uniform closure of the cords (via the edge mechanism) unites in one deft stroke our physical-emotional-intellectual self; the primitive and the refined are reconciled, and so are the mind and heart. The mechanisms of emotional expression are fully incorporated, and the tongue is freed to form words. Individuality is released in sound through virtue not of a different way of using the vocal structure but by the different length and bulk of our cords, the rest of our physical make up and the influence of our own unique experience of being and living.

Before words

The journey beyond, behind or before words may be disconcerting. It leaves the voice on its own, independent and seemingly pointless. No longer a vehicle for words and music, it is, nevertheless, a whole barely-explored world of eloquent sound in its own right. The significance of this must not be underestimated or denied, for it expresses our existence without the necessity to explain it, let alone apologise for or boast about it. Left to its own devices, the wordless voice may feel threatening, since it strips the singer of protective clothing, masks or pretence. It sounds the singer out, showing her for who or what she is: characteristically and deeply human, utterly unique and therefore utterly alone. But for the fact that we share this 'plight', our uniqueness might be untenable. Singing shows us, however, that we are part of a deeper and broader scheme of things, if only we can connect to it.

In singing to people with honesty we make an unconditional connection of the spirit. The 'shared breath' between singer and listener vibrates with the life of the singer and what she is singing about. As singers we acknowledge our participation in the world by affirming ourselves in public, by filling the space with the personal and universal content of our sound. At the same moment the listening world is included through an involuntary act of resonation.

Restoring our voice to its natural state can be seen as journeying into our human existence and shared evolution. At base the process is manifestly physical, employing the full extent of our breathing apparatus which includes the powerful muscles that keep our spine in shape, while the primary impulse for giving voice is essentially emotional. This system provides us with the perfect physical-vocal apparatus for our intellect. In spite of this arrangement, singers generally seem to veer either towards a highly physical or a somewhat intellectual approach to their singing. Neither extreme is unifying or fully satisfying because it is unbalanced, the one denying the voice's fulfilment in intelligence and the other denying its roots or physical-emotional grounding. We might deduce from this that the emotional element (which is often more evident, albeit crudely, in singing with a physical bias) is the unifying factor when it comes to singing, except that clear and direct expressions of feelings in sound are entirely dependent on physical ability. Singing as an artistic activity develops at least semi-consciously from the need to express ourselves lyrically as well as intellectually and emotionally.

The three basic human elements in singing – body, emotion and mind – are tempered and made effective by a balanced structure, with no one element over-ruling or overriding the others. A bias towards one element can be made according to its weight in whatever is being expressed. Similarly, when a true vocal response is assured, a vocal phrase can be instigated by the heart, the mind or the body. Balanced, the voice is capable of great variety. Out of balance it can easily become something less than the singing voice. Prolonged vocal specialisation, for whatever reason, puts its capacity for communication in jeopardy.

'Going deeper' is not just a matter of will. The desire to connect on a deep level whatever the condition of the singer's voice may create a more compelling communication, but where the voice doesn't respond to intention it will be forced or effortful. Modern performances have tended to disregard the human communicating imperative, preferring instead accurate reproductions of 'the notes on the page', which require less personal commitment and are 'safer' for recording purposes. This largely ignores what singing on the human level really is. To be effective, the will to connect must be matched by the physical-emotional ability. If the listener is truly to hear and be convinced about vocal communication its source must be accessed without impediment. The artless singing of the *contadina* might be more effective in terms of human communication than singing which is schooled but not free, simply because of its honesty.

Musical authenticity and the singing voice

'Authenticity', manifested in much 20th century performance practice, has played a role in divorcing the voice from its source. George Santayana writes:

> Correctness . . . in respect to familiar objects is almost indispensable, because its absence would cause a disappointment and dissatisfaction incompatible with enjoyment . . . But fidelity is a merit only because it is in this way a factor in our pleasure. It stands on a level with all other ingredients of effect. When a man raises it to a solitary pre-eminence and becomes incapable of appreciating anything else, he betrays the decay of aesthetic capacity. The scientific habit in him inhibits the artistic.

Later he writes:

> The fact that resemblance is a source of satisfaction justifies the critic in demanding it, while the aesthetic insufficiency of such veracity shows the different value of truth in science and in art. Science is the response to the demand for information, and in it we ask for the whole truth and nothing but the truth. Art is the response to the demand for entertainment, for the stimulation of our senses and imagination, and truth enters into it only as it subserves these ends.[20.1]

The priority given to accuracy and authenticity has, I believe, skewed our perspective on singing, making us forget two fundamental facts: music that touches us is based on our feelings, and the singing voice is not a 'period instrument', but has always been with us in the human body. The singing voice and transmission of feelings are interdependent. If we insist on singing music in a certain way, with a certain sound, we may be defeating the purpose of conveying feelings or even of creating a 'meaningful' sound. So long as we see singing in terms of 'styles' we run the risk of never allowing the voice to be what it is by nature. We will continue turning out stereotyped voices in all so-called voice categories.

The diminishing power of singing

It seems that the composers from Monteverdi on for two and a half to three centuries composed their music in recognition of what the human voice could do and express as a whole. *Bel canto*, as an ideal, not only satisfied an aesthetic, but also meant 'skilful', 'expressive' and, I believe, healthy singing.

Vocal music of the 20th century (with some notable exceptions) became vocally brutalising, divisive and effortful. The singing world ignored E. F. Schumacher's maxim 'small is beautiful'.[20.2] Small, for more than half a century, has been anathema, and uneconomical besides. The irony is that the 'big is beautiful' idea has spawned generations of less than beautiful, strained singing, while producing pitifully few voices able to sing the music of Verdi or Wagner.

In the same period of time much of the singing of so-called Early Music has gone to the other extreme of passionless propriety, often arising from or creating limited vocal resources and invariably based in 'authentic' artifice. In between we have an overly 'intellectual' lieder tradition, based in part I suspect on poor imitation of, for example, the superb young Fischer-Dieskau.

Amplification

Voices which for one reason or another cannot be heard well enough are more and more frequently amplified. This is an insidious and worrying trend which, without anyone intending it to, suggests that either the proper balance of musical forces is no longer important, or that a well-balanced voice does not matter. This does little to encourage singers to work thoroughly on their voices, threatens to belittle or actually limit what they can do, and encourages larger, often unsuitable, venues to stay open, and large-scale 'commercial' ventures in super-large venues. Along with these trends the wonders of technology and commerce have standardised and packaged recorded singing as though it were some kind of European edible commodity.

While there are enough exceptions to the negative picture to make it seem not so bad (Early Music, for example, is beginning to be sung with a fuller use of voice and more genuine feeling), the general picture to my way of thinking and hearing is neither healthy nor promising. The irony is that a finely tuned voice which is strong comprises all that is necessary for drama, subtlety, flexibility, emotions and projection, whether in the singing of Early Music, lieder or opera, (given that some voices are inherently more 'dramatic') all of which require passion as well as accuracy. Contemporary music should benefit just as much from a well-balanced instrument which is versatile and flexible whether the music is written to be sung lyrically or not.

Conclusions

The oft-used quote 'only connect'[20.3] sounds as though connecting should be easy. It may in fact be simpler than we let ourselves think or imagine. The difficulty seems often to lie in imagined consequences or in a visceral wariness. Dare to connect! Dare to feel, to be a golden eagle! In daring to be yourself, you encourage others to be themselves, by sharing the solitude of your uniqueness. Through this uniqueness you reach the universal – unity in diversity. To communicate (connect) on a deep level, therefore, you must want to bring out and foster your individuality and be master of yourself. Being authentic, and true to yourself, you give hope to other peoples' aspirations.

Beauty produces in the listener aesthetic emotion to which she or he reacts. Giving hope to others through aesthetics (the privilege of artists) is an ethical act. Aesthetic performance, therefore, has ethical consequences. You dare to stand before an audience with the power to move hearts and minds. It makes no difference how good a performance looks or sounds – if you are not truly present in it no real connection is made. If you spin a line when you really want to inscribe a circle, you are reinforcing the sense of disconnectedness that prevails in the absence of true communication. This is the same as the common experience of feeling alone in a crowd. There is responsibility in knowing that you make or can make a profound difference to the life, feelings and thoughts of those for whom you sing. Being true to yourself means owning your 'message', the expressive objective which resides inside you. Your responsibility as an artist is therefore as much to yourself as to your audience. It's not a burden – it's enlivening! Being faithful to yourself means handling your freedom wisely. Knowing you are connected entails understanding the impact of what you do.

Ethical action asks for contemplation: considering the effect or the likely effect of what you are doing. Knowing where you are in the effectiveness of your communication can be obscured by the big aesthetic, the 'grand show', the sugar decorations on the cake. Ultimately, whether your voice is naturally big or small, what makes it telling, real, significant and beautiful is its connection with its source, and your commitment to this connection.

PART III

Chapter 21

The Ego and the Egoist

With good self-esteem, you can be fully present in your singing. Without it you resist exposure by withholding or 'covering up'. A liberated voice helps you to off-load the luggage of self-consciousness and fear of exposure because it is grounded, integrated, and intentionally based on connection with yourself and others. Healthy self-esteem is important to a performer in terms of clarity, authority and communication. This accords with the truism that you cannot give what you haven't got. In assessing any performance including our own we need to consider its values. A performance motivated by real and deeply considered values will be far more profound in its effect than one with only superficial intentions. Values are expressed as much in the audible aspect of a singer's performance as in the visual.

This is important because of the potential problem of what it means to 'sell' a performance. Self-effacement, timidity or apology will not produce a good performance but neither will 'showing off'. Singers must 'sell' their performance in the sense of knowing the human value of what they're doing and of what they want to say, believing in it and unashamedly sharing it. 'Selling' a performance means opening oneself up and letting people hear, see and enjoy the 'personality' in the performance.

Fear and incapacity often render impossible the ability to truly share. They seem to call on the ego for help. However, the word 'ego' is often used incorrectly. Ego can mean self-esteem, and the part of the mind developed by an awareness of social standards. 'Egoistical' on the other hand means 'self-centered, narcissistic, self-indulgent, self-aggrandizing'. An 'ego-trip' is 'An activity etc devoted entirely to one's own interests or feelings.'[21.1] Ego-directed communication is self-gratification or mere show. The singer who wants to inflate himself produces an insincere performance. This, like so many of today's products, has little depth, less durability, and about the same nutritional value for the heart and soul as most fast food has for the body. The vibes we give out, the chemistry we catalyse, our body language, active or passive, all communicate. Clearly the free voice doesn't automatically make us good communicators. The means is one thing, the desire and disposition quite another.

We might say that Art is motivated by a need from the artist to express or create, implying that it is necessarily selfish. Creation can certainly be therapeutic for the artist. Indeed, if this were not so I believe it would be of little use to anyone else. The self-motivation in the creator signals and inspires the creativity within others. Artists are often wounded people; their therapy of performance can also be ours. Creativity has to be of the individual in order to appeal to individuals. The question of values remains, and it's these for which, I believe, any self-respecting artist will search. We have as much right to indulge in this process of self-discovery as we have to be fully ourselves. The difference between this self-searching and behaving egoistically is that in the former we mine for real diamonds, while in the latter we adorn ourselves with fake ones.

Singers have to be aware of the difference between social and performing skills and to not let their social persona ('the nice little girl', 'the charming young man', 'the intelligent woman', 'the strong man') take away from or intrude upon their performance. The adoption of a persona can be a kind of self-indulgence, a misplaced bid for acceptance or applause. Singing and acting for applause or personal admiration cheapens your performance. Making contact with your audience by introducing your songs, for example, may add to an audience's comfort or appreciation, but not if it's done in an ingratiating fashion. An honest, fully committed performance holds audience's attention by virtue of its connecting power: our concern for those apart from ourselves for whom we're singing. The deeper or more genuine the concern or intention on the part of the performer, the deeper and more genuine the response it is likely to elicit. The singing voice, whether in singing or listening, is so much a part of our human make-up that a singer's expressed good will will engender good will; indifference will engender indifference and so on. This is something like the Buddhist law of karma, or of cause and effect, and is as much to do with energy as it is to do with good or bad intentions. As perfoming artists, if we want to take care of ourselves we had best be generous. We receive love by giving it.

When we perform, our intentions should lead the way and our voice should follow spontaneously. This cannot happen if the voice is the object of our mind's concentration. Self-conscious singing of any kind draws attention to the singer rather than to what she's singing about. Any thought-out vocalising is by definition 'effected'. Putting on a superficial show and manufacturing effects amount to the same thing, a blocking of genuine communication.

The need for prima donnas

Arturo Toscanini's daughter Wally describes in an interview with Lanfranco Rasponi how during the 20th century singers have been increasingly subjected to the demands of stage directors, and have also become more focused on making money, rather than dedicating themselves to their art, with the sacrifices that this entails.[21.2] I have come across many singers desirous to serve their art but few prepared to make the necessary sacrifices – a strange paradox. Sacrifice, in a greedy consumerist world, where so much is provided or taken for granted, may not be a popular concept. Young professionals often discover that they are ill-prepared for the demands both of the profession as a business and job of work, and of the art that makes it so special. The relationship between a singer and agents, conductors and stage directors who paradoxically may not always have their singers' best interests at heart can further complicate the situation. It is unfortunate that 'prima donna' has become synonymous with 'difficult'. While there are always exceptions, I have no doubt that, most of the time, when singers appear to be difficult or demanding in the eyes of their employers, it is because they are unduly stressed or merely being reasonable in the circumstances.

Competent singers abound, and it may be that in protesting about non-conducive conditions and unreasonable demands a singer runs the risk of disfavour or replacement. However, the risk to a singer's real progress when she sacrifices her talent to unscrupulous directors or entrepreneurs is great, for two reasons. First, given poor conditions she is unlikely to sing well and thus risks her reputation. Secondly, any singer performing in conditions in which she must distort, force her voice or sing for too long at a time risks vocal damage or arrested progress. Many stage directors and even some conductors are not interested in *bel canto* in the sense of either beautiful or healthy singing, and much abuse of singers comes in the form of their being conveniently 'used'. Two examples are budding soloists being obliged to also sing chorus in productions, and college students or recent graduates being given unsuitable roles, effectively enabling the better singers to star while they struggle. Apart from singing teachers and the medical profession, the people best qualified to educate others on how singers should be treated are singers themselves. As long as singers fight shy of standing up for themselves or saying 'no', they will continue to be treated as easily dispensible commodities.

Stage directors

Stage directors should be concerned about how singers use their bodies on stage. Although bodies and voices are meant to work as one, this is rarely the case. Before we can intelligently consider movement with regard to singing, we must understand that the essence of posture is relational rather than positional. In training, the goal is flexible alignment, but, while a singer is gaining this, he's almost bound to be as limited in his bodily freedom as he is in that of his voice. There are always individual limits that a good director must be able to gauge and discuss with the singer if he's to sing well or avoid counterproductive practice.

A singer's performance can be transformed by movement which distracts her from her voice and helps her to embody a character. A singer whose voice restricts her bodily freedom will only make matters worse by ignoring the fact. Singers must learn to gauge for themselves what helps or impedes them, but also be prepared to try out reasonable moves, positions or gestures, or else to suggest more comfortable ones. Many stage directors come into opera from a theatrical background, with frighteningly little knowledge of music and even less of singing. This is potentially dangerous to each individual singer and to productions as a whole. Not only is the drama not understood from the music, but singers are unrealistically expected to work in the same way (including unreasonable hours and lengthy periods of repetition) as actors in straight theatre. Many conductors have equally little understanding of singers and what enables them to produce their best vocal performance. Singers cannot successfully be treated as instrumentalists.

Survival strategy

Singers are so often in incompetent or unsympathetic hands that it's vital that their training lays a good foundation for their vocalising. It's a sad fact that many singers are considered to be poor actors (as well as poor musicians), simply because their bodies are locked into muscular compensation or their minds and bodies occupied with tricks of vocal technique. In order to survive their very demanding profession, singers must build up vocal stamina, so as to be able physically to stay the course. They must also develop a 'thick skin' in order to be able to cope psychologically in any environment. They must be more than well prepared for rehearsals. These are all things that with forewarning and sufficient time an aspiring professional can take care of.

***Fächer* (pigeon-holes) and tessitura**

There are few devices in vocal pedagogy more damning than the attempt to put singers in categories according to 'type'. In regenerating and thereby 'discovering' a voice, labels can be substitutes for thought and imagination, and the responsibility attached to training. Categorising a voice is convenient for those who have no ears to hear or patience to fathom its real capacities, or who are not prepared to put in the often convoluted, tricky or downright frustrating work of thorough investigation and development usually needed to bring a voice to fruition.

Hearing and working with hundreds of singers has convinced me that there are as many 'types' of voice as there are singers, and therefore every conceivable variety of timbre, range and 'dramatic' ability between voices within one gender and between the two. Sopranos with a high coloratura facility can have a large low range, so-called mezzos can sound soprano-like, tenors can have baritonal qualities and vice versa. Believing that male and female voices need to be trained differently can be an unnecessary complication. One of the problems behind premature categorisation is that voices are often judged on what they sound like to begin with, rather than on their capabilities when fully released. Imbalance born of subjective preference criteria must be addressed, if only for the sake of health.

It should be remembered that, when structurally well-coordinated, a voice stabilises to a mean pitch (tessitura): the level at which it can most comfortably sustain itself. What is comfortable for it at the outset of training may give little indication of where a voice may eventually settle. Timbre and range can also be misleading at the beginning.

Regarding repertoire

A singer should always choose what is within his or her voice's present capability (the extent to which it can sing flexibly and without strain), or else what feeds its ongoing process of liberation and development. Composers would do well to ask singers for whom they're writing not only the limits of their range, but their comfortable tessitura, so as to avoid the common mistake of writing music which is either constantly low or constantly high.

It should be remembered that it is when the voice is best balanced that it is at its most flexible and able to reach its natural degree of dramatic ability, with the greatest amount of effective tension in the folds. It is only when a voice's natural tessitura is achieved that words can be expected to be clear throughout its natural range. If beauty is part of the aim, remember that it is only by regenerating a voice's natural capacity that this can be achieved objectively. Because this quest can take years, it may appear that a voice develops into something different from what it was originally. This is a deception, for we cannot alter the constitution of a pair of vocal folds, but only help them to gain their original potential. All attempts at making a voice something that it is not by nature will always prove less than satisfying and are bound to fail sooner or later. A voice handled with its human, individual and instrumental nature in mind will, in one way or another, always be coming into its own. Employers would do well to recognise that most young professional singers are still in the early stages of their vocal development.

Specialisation

When music is introduced into the primarily vocal process, the teacher must always take into consideration the state of the voice as such, and only then consider matters of style and interpretation. Attempting to be 'authentic' too soon usually prevents singers doing as well as they can (in accordance with what their voices will do, given freedom rather than restrictions). It can hold up the process of development, lead to premature specialisation, and even persuade singers that, to take one example, Early Music is the easiest and opera the most demanding genre of vocal music to sing. Premature specialisation is tantamount to a vocal sentence, effectively arresting the liberating process.

One of the most damning aspects of specialisation in general is that all styles have suffered because of it. Many highly musical or literate singers have confined themselves to Early or Contemporary Music. They may have excelled in these exclusive arenas, but have been unable to fully express themselves. While this might have been the individual's choice, the music itself has often suffered from lack of vocal drama, emotion, sexual energy and other qualities that characterise the complete voice. Opera singing frequently lacks the finesse which a more thoroughly developed voice could bring to it.

Confidence is built by being able to sing, not by forcing yourself to sing things for which as yet your voice isn't ready. Getting through an aria is one thing, making a good or even satisfactory job of it quite another. The latter feeds confidence because instead of exposing limits it demonstrates true worth. This, achieved with committed awareness, rewards you with courage to go further. Real progress is to be found and measured in the ease, not the difficulty, with which you accomplish your singing.

PART IV

Back and Beyond: Redefining *Bel Canto*

The Spirit of Wisdom and Diversity in Unity

PART IV

CHAPTER 22

Healthy Communication

We humans are unique among animals in our inability to live and express ourselves naturally. No other creature permits itself to exist in such contradiction to nature. Although our civilisation appreciates singing and has evolved various cultural vehicles for it, it is evident that the civilising process has been responsible for serious generic vocal deterioration. It might be argued that a civilising process is needed to ensure peace and harmony as well as survival on an otherwise hostile planet.

People across the world are discovering that Nature will only tolerate so much abuse. Having failed to see what was going on or else denied it, we are now paying for centuries of increasing pillage, pollution and manipulation. It seems that a lot has to go wrong before we admit that something is not right. Surely by now we should realise that there's more to good health than simply not being ill. If not being ill doesn't necessarily mean we are well, what can it be that makes the difference between apparent and real health?

When it comes to communication, we meet a striking paradox. One might expect communication in the sense of understanding one another or relating socially to be a major objective of all human beings, especially in the pursuit of peace and harmony. We all possess not only hearts and minds, but a voice to demonstrate that they are fundamentally one. Human intelligence, which speech has played a major role in bringing about, has produced amazing linguistic variations and sophistication. The human voice facilitated this but has also suffered speech's fragmenting effects.

When we are born we seek the sound of the voice that 'speaks' to and of us. As the cries of delight and protest die down, however, the voice's potent, spirited dynamics gradually fall into disuse or enfeebled use and the sound image of ourselves fades or becomes distorted.

The mind effectively ends up ruling the heart and causing division which is not only 'explained' and emphasised in words, but demonstrated day in, day out in the sound of our *conditioned* voice. We become in effect a house divided against itself. This is not something that we like to acknowledge. On the contrary, fear of alienation keeps us from being truly honest and open. Our voice betrays us as much as we betray it, and we get so used to this that we don't think twice about our everyday inarticulateness.

Intelligence and emotion (mind and heart) are shared by all human beings. We cry out our anguish or anger or whoop our happiness in the same way with the same voice whatever words or explanations we might attach to experiences. The voice we are born with has the capacity to join us to ourselves and communicate us, being and spirit, to others with its own intelligence. When this capacity is diminished or distorted we should worry for the spirit of humanity at large.

Even if we were to learn every language under the sun, we would find words inadequate. It would be difficult to imagine a world without singing – even unhealthy vocalisation that passes for singing shows the desire and need to communicate the great variety of human feelings and experiences with the intelligence of the mind and the heart harmonised in sound.

Our mind transforms this need into high art, encouraging us to seek out our most beautiful and versatile sound. Our lyric disposition, like our feeling person, seeks a way of expressing itself in sound. It is interesting that it is the singing voice which is best able to satisfy these human imperatives. This voice reminds us that speaking and emotion are not necessarily at odds, but are rooted in one desire for reciprocal communication. The singing voice tells us that both can find a form of expression which is universally 'intelligible', acceptable, informing and reassuring.

Our fundamentally *human* (as distinct from cultural) sound is the one with which we emote, speak and sing music simultaneously, all these means of expression coming from the same source, physically interdependent. Surely these cannot have been intended to exist in anything but a harmonious state? When comparing the liberated singing voice with the normal speaking voice it is clear that the difference is not in quantity or strength but in its *functional extent* and overall physical-emotional integrity.

Examined in this way, we can only conclude that the normal person's voice is an organ in decline and that in terms of personal communication it has brought us to a critical point of collapse and inarticulateness, and close to emotional bankruptcy. The irony is that it was with speech that we became *homo sapiens*. What makes the crucial difference vocally, both in competence and depth, is what provides the centre point, or the missing link, so to speak, between the intelligent body and the intelligent mind: those astonishing muscle fibres at the leading edges of our sound.

I have no doubt that, when it comes to expression in sound, we can experience and 'know' a person more intimately and thoroughly through his 'singing' voice than through his speaking voice. In our privileged position as singers, in discovering the core of our voice (an inner search), we are enabled to voice our heart and mind openly and harmoniously. This is a matter of accessing and utilising a force lying at the core of our being which is desirous to reach out – the communicating imperative.

It is through sharing our individual experience and our collective humanity (two common viewpoints) that understanding and companionship can be achieved; sharing, not imposing or competing (which are both divisive). It is in integrating our basic (primal, instinctive) self with our refined, more intellectual self that we gain the self-confidence to do this.

In training, resistance to change is greatest at the visceral ('being') level. This raises the whole question of the difference between the absence of real or imagined pain and the presence of full rich life, and between existing (coping with our condition) and improving, creating and enjoying living. I believe that singing, with its connecting capacity and generosity of spirit, teaches us how to be fully alive. Healthy singing isn't merely an absence of dis-ease in the throat but something a great deal more positive. Singing embodies the essentials and true spirit of creative being: physical health, feelings fully shared, intellect plainly articulated.

To find our voice we have to be re-creative. We must recombine our various 'parts' holistically so that they give us a creative capacity beyond or above their sum. You may point to people who have created marvellous things (physically and intellectually) in performance, but wonder what they would have created had they fully embodied and lived their creation.

In a sense we *return* to our wholeness, or to our potential – we certainly do not achieve it by adding anything. Perhaps it's even a mistake to endeavour to achieve something special. The singing voice is merely something we are born with.

To quote T. S. Eliot:

> *We shall not cease from exploration*
> *And the end of all our exploring*
> *Will be to arrive where we started*
> *And know the place for the first time.*
> (from 'Little Gidding' from *Four Quartets*)[22.1]

To live is simply to be who you are. Since this is not generally what we are encouraged to be, we have to discover what it means. Singing is a good way to do this. The concept of singing as 'special' is fallaciously based on the fact that so few people can sing. When a voice is whole, however, we as teachers find ourselves saying to singers, 'don't try so hard', 'let the expression be natural', 'just "speak" your singing', as though it were the simplest, most natural thing in the world. *Having* a voice, or the wherewithal to sing, as though it is something we've gained or been given, is quite different from *being* a voice, being what it is that naturally expresses *you*. An audience may respond to what words and music on their own evoke, but the singing voice engenders an altogether different response, because of its quality of personal presence.

A performance is 'sold' through virtue of being vitally in the moment. This quality of being present or of present being is the mark of healthy communication. The desire to communicate and the desire to receive are inevitably and fully satisfied. Vocal health can be measured by these criteria alone. A liberated voice will not just entertain, distract or inform a listener, but will wake him from the inside to what is being communicated on his behalf. The listener is moved empathically by what is alive for him in the singer's communication.

PART IV

Chapter 23

Redefining Bel Canto

Nowadays it seems to be popular to dismiss the past and assume that what's new is automatically an improvement. Naturally if we cling to the past, or think that things will 'never be the same', we may be equally naïve or blinkered. What I believe is good from the past is what can be built upon or developed while informing a creative life and sustaining human values. What is interesting or beneficial for the future is what contributes to better living now. Lasting values are those that have foundation and make for health of the individual, societies and the world at large.

Whatever we think of the music associated with *bel canto*, we cannot avoid seeing at a glance that it was not simply a style to be sung in a certain manner. Less still was it merely a vehicle for vocal gymnastics or a beautiful sound. It is widely believed that as a manifestation of 'excellence' in singing it was unsurpassable. The singers of the indisputably taxing music of Rossini, for example, would have come from a background of music originating from the Italian Baroque masters, whose vocal writing already demanded precision, agility, and long lines. Rossini and his contemporaries simply stretched their singers to the limit. Vocal music since has debatably only been more difficult when it has strayed from the 'lyrical', and gone against the nature of the singing voice, tempting or obliging singers to strain against the odds.

To imagine that at some point in history somebody decided to encourage singers to sing beautifully (*bel canto*) in a purely aesthetic sense is as alien and irrelevant to the nature of singing and to the principles of expressive, skilful vocalisation as to imagine that beautiful singing no longer matters. Both ideas are 'anti-singing', the one because it shows a lack of understanding of the nature of *bel canto*, the other because it devalues it. A serious sticking point seems to be the inability or unwillingness to recognise or accept 'beauty' as a natural outcome of good singing rather than an aim. Beauty of sound is only one of many attributes of the liberated voice.

We can safely assume that, castrati apart, the *bel canto* singers were no more of a special breed than singers are today, but judging by what they were expected to sing, they must have had superior flexibility, not only in terms of agility (which, significantly, was expected of all voices), but in terms of 'colouring', dynamics and emotional expression. These among other skills were the natural consequence of well-balanced vocal integration. Everything written in the heyday of *bel canto* singing could be reasonably expected to be performed by a well-trained singer.

Since the late 19th century, singers have had to cope with more and more non-lyrical music, unsympathetic orchestrations, works written by composers who have not considered the nature of the singing voice, and larger and more powerful orchestras. The effect has been disastrous for singing! Singers are no longer encouraged to sing with integrated vocal and expressive skills, the thing that best defines *bel canto*. What is the use of being able to sing 'big' without flexibility, or accurately without emotion? Who wants to sing ugly?

'Big' and 'Belting'

Bel canto can be erroneously linked with 'overblown singing'. Certainly the music of composers such as Bellini could be grand and florid, but I doubt if it ever received the exaggerated emotional treatment that often seems to be the lot of later composers. Only in the last century has 'big' been reckoned to be beautiful. This has led in one direction to the more modern style of 'belting', which paradoxically needs a microphone to make it carry. Incidentally, 'belting' is not grounded singing; it is unbalanced in favour of chest voice, precludes mixing, and is the antithesis of subtlety. It therefore communicates mainly on a crude level.

Losing sight of 'beautiful singing'

Whilst the post *bel canto* giants of expressive innovation, such as Verdi, Puccini and Wagner, must have written for singers brought up in the *bel canto* tradition, I don't believe there is any reason to blame their music for singing which subsequently favoured 'fatness' of tone, volume or heaviness over intensity, the obvious over the subtle, 'emotional effects' or false emotionalism over genuine feeling. Larger performing spaces, acoustics unsuitable (over-reverberant) for singing and greater orchestral volume must have encouraged such excesses, which have turned out to be self-defeating. Deterioration of vocal excellence and prowess must also have been exacerbated by vocally fragmenting, harsh and angular compositional experiments in the early 20th century, which were the result no doubt in part of the aggression, pain and heartache of two world wars and various totalitarian regimes.

In *bel canto* times, the combination of an 'emotive' tone and musical fluency must have resulted in magically uncomplicated but moving performances. The deliberate 'putting on' of emotion is as alien to the singing voice as its suppression. It is interesting to note that the overtly emotional *verismo* style, short-lived as it was, coincided with the beginnings of visual investigations into how the voice worked, and the beginnings of systematic investigations into the human psyche. Freud (often seen as the father of modern psychology), Manuel Garcia Junior (the inventor of the laryngoscope), and Schoenberg (a major pioneer of atonal music) were all born within a few years of each other. While one might have thought that psychoanalysis, scientific vocal knowledge and a new musical aesthetic might happily have combined to make something even greater of singing, it seems that instead they unwittingly conspired to pull the voice apart.

Only in liberating our singing voice and proceeding beyond the point of conscious vocalisation do we discover what our voice is really capable of in musical and expressive terms. As this book has shown, I am interested in what *more* can be achieved, not by futile attempts to add to or interfere with Nature, but by imagining the human voice, at least temporarily, free from the dictates and unnatural constraints of style and commerce.

Once we've established that *bel canto* is not limited by its literal translation and isn't a style of singing confined to a certain period, we have freed ourselves to imagine its further possibilities. I have implied that the standard of singing could be higher if the capacity of the human voice was not generally so underestimated. What was achieved by the old Italian singing teachers was a solid base of competence and integrity, to which all would-be singers with sufficient talent could aspire, given time. I suggest that, given such a base, singing in our time could be taken to a deeper or higher expressive level. This would be the outcome not of creating 'super voices' but of greater awareness of the universal nature of the singing voice, which, although it has been neglected or degraded to some extent, hasn't changed any more than our larynx or breathing system has. What has been developing ever since the birth of modern psychology is our *consciousness* regarding ourselves. 'Going back' to the vocal ideals of the *bel canto* school would not only be an advance, but would create the technical possibility for progress.

Self and other

In regenerating your voice you learn to connect and communicate with yourself in such a way that you don't jeopardise your own integrity when it comes to communicating (committing) to others. Communication (relationship) is all the more powerful when the self is preserved intact. How can one be clear and coherent in relating, in committed communication, while maintaining respect both for the other and for oneself? How can quality, clarity, rigour and durability be reconciled in this commitment with necessary changes, evolution and mutation? I believe that freedom of voice facilitates 'being in sound', and the ability to share this existence with other beings. This state of 'truly being' resonates in an individual and generally. Thus a singer reaches every individual in the crowd and by extension the crowd *en masse*. (*Note:* This power should not be confused with the dictator's rhetoric, which by threatening, belittling or devaluing the individual's status or integrity arouses a primitive or hysterical herd instinct!)

Learning to sing mirrors progress in life, a synthesis of realisable personal attributes, experiences, relationships, followed by new growth. The birth of modern Western psychology brought with it the promise of more conscious processes of development and synthesis. In spite of Carl Jung's concept of 'individuation' (individual becoming),[23.1] I believe psychology has been dogged by a failure to treat the person as flesh and blood as well as mind. We now have physical therapies, however, which treat the body and mind as one, and there is an increasing inclination towards holistic treatments in health generally. These developments and similar ones in other areas of life are gradually helping to create a holistic view of existence unprecedented in the West.

As I see it, the study of singing is still far from being holistic because, although physical, mental and emotional aspects are attended to, they are still too separately delineated to result in the kind of synthesis which leads inevitably and organically to new growth. The term 'physical' implies physical skill, 'mental' implies academic comprehension and emotion implies at best something optional. The true value of what we are doing will only be appreciated if we aspire to *the whole*. All aspects of the voice acquire their true value through mutual support and a kind of 'alchemical becoming'. Only when we tune into this whole nature will we confidently be able to assume it.

Like culture in general, singing satisfies a deep need to relate, to consolidate and to celebrate. It is perhaps healthy therefore to see creativity not only as an 'appreciation' of life, or even as an instinct, but as the imperative to share, or the capacity to give. In waking up to our potential we give value to ourselves, and, through consciousness, we give value to the world. Through singing and art we not only find our individual voices but tap into the 'collective voice' of humanity. This is the ultimate power of the singing voice, a power not imposed but shared, effectively empowering others through resonation.

Collective unconscious

A friend of mine once said, 'evil is not the opposite of good but its absence'. Even if we don't have evil thoughts and even if we don't do evil things we can think and imagine doing evil. If we don't think or do evil it's because 'we know better', and we know better because we feel better. In becoming fully human we recognise good, desire it and become part of it. To become fully human, however, we need favourable conditions, like plants that flourish on light, air, water, and sunshine.

People often glibly say that human nature cannot change, implying that we have to live with the rough and the bad as well as the smooth and the good. My own view is that we human beings are still growing up, fathoming and realising our nature, and not so much changing as still coming into being.

As a baby entering the world we inherit 30,000 generations of human experience. This is our shared humanity, the collective unconscious. Exploring this part of our being may be painful, since our own individual experience is mixed in with it, but doing so is a means of understanding what needs to be done for good – how to deal with suffering and perhaps turn it to our own and everyone else's advantage.

Singing celebrates our pain and joy as human beings through text, drama and characterisation. We can imagine and temporarily act out characters' deeds because, through our human lineage, we share them. Beautiful music and vocal lines go into the ugliest or most anguished themes, the Passion of Christ being one example. The more we appreciate expressions of the human experience, the more the burden and the celebration can be shared and the more understanding and compassion can be generated.

Our visual, technological space age has encouraged us to look outwards instead of inwards, to see what 'out there' has to offer, how it can 'support' us and make us feel better. In the process, I believe we have become less significant to ourselves and to everyone else. The opinion of the director Richard Eyre is that 'the only unexplored territory now is the territory of the human soul. Instead of gazing at an unknown continent, we peer inwards to a landscape that is often as mysterious and inaccessible.'[23.2]

The singing voice is a stirring of the heart to appreciate what life is or can be. Thus we celebrate whatever moves us. Much of this is depicted in the sheer sound of our voice, just as the meaning and expression of a picture is shape and colour. Even Goethe went so far as to say, 'our best convictions cannot be translated into words. Language is not good for everything.'[23.3] The voice probes and penetrates the deeper, seemingly inaccessible levels of our being: the inexplicable, the irrational, the heartfelt and the soulful.

Gender

It is gradually being recognised that a major symptom of our human immaturity is the assumed inequality and lack of understanding between men and women. In Jungian psychology our 'male' and 'female' characteristics represent mutually inclusive sides of the human character, or characteristics which can be mutually completing and fulfilling. Given specific biological imperatives, this means not only equal status between female and male but also the balance of characteristics in all individuals regardless of their sex. The perceived male-female divide damages common consciousness which can only come into focus with integration.

The need for equality and mutual inclusion is clearly demonstrated in the singing voice where balance (and therefore full potency) is only achieved through successfully marrying the 'female' and 'male' elements. The need for a put-on image melts as these facets, essential to being fully human, merge and potentialise one another.

Perfection

The balanced singing voice, for all its skill and power to move, is no more 'perfect' than a well-balanced posture, which enables a person to use his body effectively and economically. Aspiring to this balance, however, is aspiring to being oneself, with the added possibility of exploring and bringing to recognition the emotional and soulful landscape of the human experience at large. The potential of unconditioned personhood is thus directly and unconditionally aired.

Connecting with people in real life (at least on any consequential level) seems to be a diminishing talent, if it ever was one. Singing, with its depth and directness, can be a stark and startling reminder of how closely we can or could communicate with each other if only we 'found the voice to do it'.

PART IV

Chapter 24

'Beauty is Truth, Truth Beauty'[24.1]

Looking for the beautiful

'Beauty is in the eye of the beholder', we sometimes say. If it were that simple anything might be seen as beautiful. Indeed, when art gives up on aesthetic qualities or values and begins to embrace ugliness we might begin to wonder if there's anything more to the concept of beauty than individual taste.

Inasmuch as Art is an expression or celebration of life, and not something necessarily useful, didactic or decorative, it clearly satisfies an important human need. Subjective ideas about beauty are not valid criteria by which to appraise a singing voice, since it is a phenomenon arising from a person's 'truth'. Our voice-freeing, as much as it's a rediscovery of natural balance, is a search for this truth – a personal journey with a personal outcome. It seems desirable and useful therefore to consider the difference between the ethical and the aesthetic. The ethical is the spirit longing for goodness, justice, truth and so on. The aesthetic is to do with the longing of the senses and might be seen as concerning the material creature (the created and the creative). A satisfying definition of beauty might be found where the spiritual meets the material, where the one is recognised by the other, as in singing.

Being the truth

The experience of looking into a person's eyes candidly is powerful or disturbing, simply because it is a clear, unalloyed revelation of the truth, inviting or else challenging us to connect, to comprehend, and to believe. It's exactly the same with the voice, which can also be deep and candid, or else superficial, ambiguous or lacking focus. Like the eyes, the singing voice can invite us in, into the singer's soul and into our own. We know instantly when this contact, whether with the eyes or the voice, is genuine. We recognise instinctively when a voice 'rings true', when voice and singer are unmistakably one, and crucially and unashamedly letting you know it.

Searching for the truth is like looking for the light or seeking enlightenment. As the truth hits us or gently dawns, we see that it's not something that we *make* or *do*. It simply exists, being itself. Can we ask for more than gaining our self and finding the self of another? The light in the voice, like the light in the eyes, simply reveals itself – everything of our world inhabited and illumined at its core.

The significance of beauty

The singing voice is like a multi-faceted diamond which, if it is well-cut and polished, reflects your true nature and worth in sound. This is why training and care of the voice must include attention to and care of our inner life. In training we 'sound' our inner self, giving ourselves feedback in the process of becoming discerning listeners. This is a mutually reinforcing process, which so long as it is pursued with an open mind and heart leads inexorably towards wholeness and health. This diamond inside, in all its clarity of colours and character, is priceless, not because it is unusual but because it is unique to each of us.

The significance of truth

Only what is understood can be properly and convincingly communicated. Therefore in truthful expression we demonstrate our understanding. Strictly speaking, the unexpressed remains the non-understood. The voice itself can bring understanding because it is as much the voice of the body, the heart and the spirit as it is of the mind. By the same token it is self-nourishing. A simple example: a singer makes a singing sound and it has an immediate effect – disturbing, delighting, relieving, embarrassing – 'understood' in other words on the emotional rather than the intellectual plane. This effect may be quite unexpected but is not altogether unfamiliar. The emotion was already there, and only needed sounding or releasing to be recognised.

Re-discovering the singing voice is reconnecting to the self. The truth in singing is therefore expressed in the voice's 'I-ness', not in an ego-centric sense but in the sense of personal authenticity.

Taking care of beauty and truth

We are surrounded by ugliness in one form or another – it's practically taken for granted. It intrudes into our lives (for example in the media) and it is increasingly clear that it is being systematically (if unwittingly) caused or manufactured by 'the powers that be' in the name of, for example, economic growth. Persistent disregard for genuine, equally shared quality of life threatens to wear down the human spirit. This suggests to me that beauty and truth must not only be sought out, recognised and understood in terms of their value but must also be fought for and nurtured.

The process of looking for beauty is not simply a question of going to the park or the sea. The immensity of the ocean or the splendour of the sunset may be lost on us unless we can to some degree engage with them. Appreciating beauty is not simply a question of making a distinction between flowers and litter, but of opening our eyes and intelligence (inner seeing) to their significance. It is a question of seeing and in our context hearing, not just looking and listening, and of being fully aware of our response. In this way we not only acknowledge the existence of beauty, not only 'pay our respects' but, as the artist Samuel Palmer put it with regard to landscape, we receive it 'into the soul'. [24.2] In so doing, we come to realise that we are as much part of the beauty we are experiencing as it is of us; we are not separate entities but interrelated and interdependent. The fact that we affect it as much as it affects us accords with the holistic principle and describes perfectly our relation with our own voice and that of others. In beautiful singing – *bel canto* – we acknowledge, embrace and celebrate the human inner landscape.

The eye of the beholder may need a more penetrating, discerning or unbiased vision in order to see beauty clearly – to see its truth. As human beings, we can easily distinguish between ugliness which repels and beauty which attracts. When we are fully conscious of beauty, I think it invites us to take care of it. Take for example the apparently simple act of walking, which actually involves about 250 muscles. When examining posture and its effect on singing, we neglect this miraculous activity to its disadvantage and our own.

If we fail to see the beauty in the various activities of our body, and ignore the invitation to take care of them, we dishonour our being and limit our creative capacity. Nature responds naturally to how we respond to her. Whether we thrive or deteriorate, she responds to our choice in how we live.

Postlude

It is a significant by-product of this book that in adopting a holistic paradigm with regard to the human voice and its health we gain a deeper and broader definition of singing. This may be highly desirable at a time when almost any vocal sound employed in the service of music is called singing. For those of us who teach singing, even if our primary task is the coaching of music, clarity about what we are dealing with, both in material and more human terms, is fundamental.

However they come about, split-off facets of the whole voice betray their struggle to unite, to become whole. Whether, for example, a voice is primitive (harsh or crude), or sounds 'lost' (vague or non-committal), or 'intellectual' (worked out, stating, explaining), such facets betray within the singing human a lack or fear of completeness, a wound that the impulse to sing normally exposes.

It may be interesting here to remind ourselves that compared with other musical instruments the singing voice is relatively amorphous until it unites itself on impulse. This spontaneous unification can happen only if the voice is fit and responsive. It is this responsive whole that we should study if we want to know what the singing voice is to the human being, and indeed what the human being is to singing.

The singing voice can alleviate the conflict within ourselves, and between us and our listeners, bridging the gaps of separation, proclaiming that we can be one if only we hear or listen for inclusion, instead of remaining deaf to exclusion.

In dealing with the questions raised in this book, we have discovered that the reasons for vocal breakdown mirror the degeneration of the human being. Singing in general has become more of an effortful striving to express experiences in a formal or stylised way than a simple, natural, spontaneous gesture in response to our joy or our suffering. In holding onto the remnants of its voice, humanity is holding onto the remnants of itself. The problems for singing – singing music and communicating through it – are the problems specific to the singing *voice.* It is difficult to view singing as some self-contained activity when its difficulties are rooted in the human condition. On the other hand, it's easy to see why singing can be viewed as something exclusive when this condition has been accepted as the norm.

What kind of civilisation is it, we could ask, that allows itself to lose such a potent unifying force, unless it is one that is going deaf to itself? If there is one most important message of this book, it is 'listen to *hear*!' Humanity's need for a common voice has never been more pressing than it is today. But we can vocalise only what we hear. If we want to commune more fully with ourselves and with others, we must strive to hear comprehensively, with clarity and compassion.

Endnotes

Chapter 1: ***Sounds Of Life***

1.1 Bérard, Guy (1993) *Hearing Equals Behaviour*, New Canaan, Connecticut: Keats Publishing Inc. p. 4.
1.2 Verney, Dr Thomas and Kelly, John (1981) *The Secret Life of the Unborn Child*, London: Sphere Books 1982 (repr.) pp. 7–8.
1.3 See Hans von Leden's account (Chapter 2) in Sataloff, Robert Thayer (1997) *Professional Voice: The Science and Art of Clinical Care* (2nd ed.) San Diego: Singular Publishing Inc. p. 82.
1.4 Tulloch, Sara (ed.) (1993) *The Readers' Digest Oxford Wordfinder*, Oxford: Clarendon Press p. 258.

Chapter 2: ***Sounds Intelligible***

2.1 Onions, C. T. (1996) *The Oxford Dictionary of Etymology*, Oxford: Oxford University Press p. 930.
2.2 Sataloff, Robert Thayer (2005) *Voice Science*, San Diego: Plural Publishing Inc. p. 58.
2.3 A term taken from Husler, Frederick and Rodd-Marling, Yvonne (1976) *Singing: The Physical Nature of the Vocal Organ* (revised edition), London: Hutchinson p. 20 onwards.
2.4 Ibid. p. 2.

Chapter 3: ***The Stifled Cry***

3.1 The relationship of singing and the emotions, and the endorphins they produce, have long been topics of discussion in musical and scientific circles. For a recent discussion of this topic, see Michael Balter (2004): 'Seeking the Key to Music', *Science* Vol. 306 pp. 1120–22, and Patrick Chiu (2003): 'Music Therapy: Loud Noise or Soothing Notes?'*International Pediatrics* Vol. 18 no 4, pp. 204–8.
3.2 Stevens, Anthony (1990) *On Jung*, London: Routledge p. 84.
3.3 Quoted in Montagu, Charles (2004) 'Mind Matters', *Resurgence*, No. 224 p. 12.
3.4 Smail, David (1998) *How To Survive Without Psychotherapy*, London: Constable Publishing. Unfortunately I was unable to trace the page reference.

Chapter 4: ***Bowing To Life***

4.1 This famous quote was used in Myss, Caroline (1997) *Anatomy of the Spirit*, New York: Three Rivers Press. Unfortunately I am unable to trace the original source of this quote.

Chapter 5: ***The Inspiration Of Life***

5.1 Sataloff, Robert Thayer (1997) *Professional Voice: The Science and Art of Clinical Care* (2nd ed.) San Diego: Singular Publishing Inc. p. 115.
5.2 Ibid. p. 115.

Chapter 7: ***Making Connections***

7.1 Capra, Fritjof (1996) *The Web of Life – a new synthesis of mind and matter*, London: Flamingo p. 154.

Chapter 9: ***Vocal Liberation***

9.1 Husler, Frederick, and Rodd-Marling, Yvonne (1976) *Singing: The Physical Nature of the Vocal Organ* (revised edition), London: Hutchinson.

Chapter 10: ***Training Ground***

10.1 Husler, Frederick, and Rodd-Marling, Yvonne (1976) *Singing: The Physical Nature of the Vocal Organ* (revised edition), London: Hutchinson p. 24.
10.2 Ibid. p. 27 (note).
10.3 Murdock, Ron (1996) 'Born to Sing'. Available from URL: www.alexandercentre.com/For Musicians/The Alexander Technique For Singers (accessed 2000).

Chapter 11: ***Hearing Our Way***

11.1 I recommend the highly informative explanations of placing with appropriate vowels to be found in Husler, Frederick, and Rodd-Marling, Yvonne (1976) *Singing: The Physical Nature of the Vocal Organ* (revised edition), London: Hutchinson. An illustrative CD can be purchased with the book – see bibliography for reference.

Chapter 12: ***Trial And Error***

12.1 Fritz, Robert (1989) *The Path of Least Resistance: learning to become the creative force in your own life,* New York: Fawcett Columbine p. 116.
12. 2 Ibid. p. 165.
12.3 Zeldin, Theodore (1995) *An Intimate History of Humanity*, London: Minerva p. 441.
12.4 Holden, Robert (1993) *Laughter, The Best Medicine*, London: Thorsons p. 23.
12.5 Ibid p. 24.
12.6 Todd, Mabel (1968, repr. from 1937) *The Thinking Body*, New Jersey: Dance Horizons Inc. pp. 37–8.
12.7 Schnebly Black, Julia, and Moore, Stephen (1997) *The Rhythm Inside: Connecting Body, Mind and Spirit Through Music*, New York: Sterling Publishing Company Inc. (repr. Alfred Publishing Company 2003). Introduction by Pat Moffitt Cook p. xv.
12.8 Ibid. p. xv.
12.9 Both quotes Ibid. p. 31.
12.10 Ibid. page 68.
12.11 Tulloch, Sara (ed.) (1993) *The Readers' Digest Oxford Wordfinder*, Oxford: Clarendon Press. p. 1061.

Chapter 13: ***Muscle Training and Fitness***

13.1 Saxon, Keith G. and Schneider, Carol M. (1995) *Vocal Exercise Physiology*, San Diego: Singular Publishing Group Inc. p. 111.

13.2 Ibid. p. 47.
13.3 Ibid. p. 51.
13.4 Ibid. p. 54.
13.5 Ibid. p. 54.
13.6 Ibid. p. 56.
13.7 Ibid. p. 60.
13.8 Ibid. p. 66.
13.9 Husler, Frederick, and Rodd-Marling, Yvonne (1976) *Singing: The Physical Nature of the Vocal Organ* (revised edition), London: Hutchinson p. 4.

Chapter 14: ***Singers' Health***

14.1 Sataloff, Robert Thayer (1997) *Professional Voice: The Science and Art of Clinical Care*, San Diego: Singular Publishing Group Inc.
14.2 Saxon, Keith G, and Schneider, Carol M. (1995) *Vocal Exercise Physiology*, San Diego: Singular Publishing Group Inc. p. 123.
14.3 Ibid. p. 6.
14.4 Sataloff, Robert T. (1997) *Professional Voice* p. 336.
14.5 Saxon and Schneider (1995) *Vocal Exercise Physiology* p. 6.
14.6 Batmanghelidj, Dr Fereydoon (2003) *Water and Salt: Your Healers From Within*, Norwich, UK: Tagman Press p. 2.
14.7 Ibid. p. 182.
14.8 Ibid. pp. 160–3.
14.9 Ibid. p. 159.
14.10 Sataloff, Robert T. (1997) *Professional Voice* pp. 457–470.
14.11 Sataloff, Robert T. (1997) *Professional Voice* pp. 291–298.

Chapter 15: ***Gurus or Guides?***

15.1 Ristad, Eloise (1982) *A Soprano On Her Head*, Utah: Real People Press p. 134.
15.2 Renowned teacher, and author, with Frederick Husler, of *Singing* (see above for reference). Together with Husler, she set up a singing school in Lugano, Switzerland. I studied with her both in Switzerland and on her return to London, in the 1970s.
15.3 Storr, Anthony (1996) *Feet of Clay – A Study of Gurus*, London: Harper Collins p. 216.
15.4 Ibid. p. 221.
15.5 Buber, Martin (1965) and Friedman, Maurice (intro.) (1965) *The Knowledge of Man: Selected Essays by Martin Buber*, London: Allen and Unwin p. 27.
15.6 Storr, Anthony (1996) *Feet of Clay* p. 225.
15.7 Codrescu, Andrei (1997) Interview, *What is Enlightenment*, Issue 12, Fall/Winter p. 62.
15.8 Wheatley, Margaret (1999) *Leadership and the New Science*, San Francisco: Barrett-Kochler Publishers Inc. p. 141 (2nd ed).
15.9 Richard Hames is a corporate philosopher, composer and writer, author of the best selling *The Management Myth*.
15.10 Husler, Frederick, and Rodd-Marling, Yvonne (1976) *Singing: The Physical Nature of the Vocal Organ* (revised edition), London: Hutchinson p. 112.

Chapter 16: ***A Clean Slate (Learning)***

16.1 I was unable to track down the original source of this quote. My editor found a reference to it on the following website: www.myisraelsource.com/content/activities. (accessed May 2006).

16.2 Lorna Marshall is a director, performance consultant and teacher. She is also a writer on acting and physicality – her recent publications include *The Body Speaks* (2001) London: Methuen.

16.3 Ristad, Eloise (1982) *A Soprano On Her Head*, Utah: Real People Press p. 30.

16.4 Suzuki-roshi, Shunryu (1973) *Zen Mind, Beginner's Mind*, USA: Weatherhill: prologue. Unfortunately I am unable to supply a page reference.

16.5 Fritz, Robert (1989) *The Path of Least Resistance*, New York: Fawcett Columbine p. 143.

16.6 Feldenkrais, Moshe (1985) *The Potent Self*, New York: Harper and Row p. 111.

16.7 Conversation with the author, 1990s.

Chapter 18: *Attraction and Repulsion*

18.1 Keleman, Stanley (1986) *Emotional Anatomy – the structure of experience*, Berkeley, CA: Centre Press p. 160.

18.2 Ibid.

18.3 Story recounted to me by a colleague.

18.4 E. F. Schumacher (1911–1977): economist, and author of the best-selling *Small Is Beautiful: A Study of Economics As Though People Mattered*, New York: Harper and Row 1973.

18.5 Merleau-Ponty, Maurice (1945 repr. 2005) *The Phenomenology of Perception*, London: Routledge Classics p. 91.

18.6 Santayana, George (1896 repr. 1955) *The Sense of Beauty*, New York: Dover Publications p. 37.

18.7 These quotes are all from Eyre, Richard (1993) *Utopia and Other Places*, London: Bloomsbury p. 95.

Chapter 19: *Being Fully Prepared*

19.1 De Mello, Anthony and Stroud, Francis (ed.) (1992) *Awareness – the perils and opportunities of reality*, New York: Doubleday p. 3.

19.2 Sataloff, Robert Thayer (1997) *Professional Voice: The Science and Art of Clinical Care*, San Diego: Singular Publishing Group Inc. p. 119.

Chapter 20: *Going Deeper*

20.1 Santayana, George (1896 repr. 1955) *The Sense of Beauty*, New York: Dover Publications p. 15.

20.2 See note 18.4 for reference.

20.3 This famous phrase originates from the novel *Howard's End* by E. M. Forster, where it is used in the context of a romantic relationship. It has since been often quoted in a variety of contexts including musical ones.

Chapter 21: *The Ego and the Egoist*

21.1 Tulloch, Sara (ed.) (1993) *The Readers' Digest Oxford Wordfinder*, Oxford: Clarendon Press p. 470.

21.2 Rasponi, Lanfranco (1984) *The Last Prima Donnas*, London: Victor Gollancz p. xii.

Chapter 22: *Healthy Communication*

22.1 Eliot. T. S. (1944) *Four Quartets*, London: Faber (Faber Library edition, 1995) p. 43.

Chapter 23: *Redefining* Bel Canto

23.1 See Stevens, Anthony (1990) *On Jung*, London: Routledge for more on this topic.
23.2 Eyre, Richard (1993) *Utopia and Other Places*, London: Bloomsbury p. 23.
23.3 Quoted in Ancelet-Hustache, Jeannette (1960) *Goethe*, London: John Calder p. 8.

Chapter 24: '*Beauty is Truth, Truth Beauty*'

24.1 From John Keats, 'Ode on a Grecian Urn', in Ferguson, Margaret, Salter, Mary Jo and Stallworthy, Jon *The Norton Anthology of Poetry* (4th ed.) (1970) New York: W. W. Norton & Co p. 848.
24.2 Quoted in Lane, John (2006) 'Luminous Vision', an article on Samuel Palmer, *Resurgence* (Jan/Feb edition) No. 234 p. 43.

Bibliography

A core text used in this book, *Singing*, by Frederick Husler and Yvonne Rodd-Marling, together with an illustrative CD, is only available directly from: Tremayne Rodd, 3 Briar Walk, London, SW15 6UD: tremrodd@aol.com

Ancelet-Hustache, Jeannette (1960) *Goethe*, London: John Calder.

Balter, Michael (2004): 'Seeking the Key to Music', *Science* Vol. 306 pp. 1120–22.

Batmanghelidj, Dr Fereydoon (2003) *Water and Salt: Your Healers From Within*, Norwich, UK: Tagman Press.

Bérard, Guy (1993) *Hearing Equals Behaviour*, New Canaan, Connecticut: Keats Publishing Inc.

Buber, Martin (1965) and Friedman, Maurice (intro.) (1965) *The Knowledge of Man: Selected Essays by Martin Buber*, London: Allen and Unwin.

Capra, Fritjof (1996) *The Web of Life – a new synthesis of mind and matter*, London: Flamingo.

Chiu, Patrick (2003): 'Music Therapy: Loud Noise or Soothing Notes?' *International Pediatrics* Vol. 18 no. 4, pp. 204–8.

Chopra, Deepak (1990) *Quantum Healing – Exploring The Frontiers Of Mind/Body Healing*, London: Bantam Books.

Codrescu, Andrei (1997) Interview, *What is Enlightenment*, Issue 12, Fall/Winter.

De Mello, Anthony and Stroud, Francis (ed.) (1992) *Awareness – the perils and opportunities of reality*, New York: Doubleday.

Eliot, T. S. (1944) *Four Quartets*, London: Faber & Faber (1995 edition).

Eyre, Richard (1993) *Utopia and Other Places*, London: Bloomsbury.

Feldenkrais, Moshe (1985) *The Potent Self*, New York: Harper and Row.

Ferguson, Margaret, Salter, Mary Jo and Stallworthy, Jon *The Norton Anthology of Poetry* (4th ed.) (1970) New York: W. W. Norton & Co.

Fritz, Robert (1989) *The Path of Least Resistance*, New York: Fawcett Columbine.

Hemsley, Thomas (1998) *Singing and Imagination – a human approach to a great musical tradition*, Oxford: Oxford University Press.

Holden, Robert (1993) *Laughter, The Best Medicine*, London: Thorsons.

Husler, Frederick, and Rodd-Marling, Yvonne (1976) *Singing: The Physical Nature of the Vocal Organ* (revised edition), London: Hutchinson.

Keleman, Stanley (1986) *Emotional Anatomy – the structure of experience*, Berkeley, CA: Center Press.

Keleman, Stanley (1981) *Your Body Speaks Its Mind*, Berkeley, CA: Center Press.

Lane, John (2006) 'Luminous Vision', *Resurgence* No. 234 (Jan/Feb).

Marshall, Lorna (2001) *The Body Speaks: Performance and Expression*, London: Methuen.
Merleau-Ponty, Maurice (1945 repr. 2005) *The Phenomenology of Perception*, London: Routledge Classics.
Montagu, Charles (2004) 'Mind Matters', *Resurgence*, No. 224.
Murdock, Ron (1996) 'Born to Sing': available from URL: www.alexandercentre.com/For Musicians/The Alexander Technique For Singers.
Myss, Caroline (1997) *Anatomy of the Spirit*, New York: Three Rivers Press.
Onions, C. T. (1996) *The Oxford Dictionary of Etymology*, Oxford: Oxford University Press.
Rasponi, Lanfranco (1984) *The Last Prima Donnas*, London: Victor Gollancz.
Ristad, Eloise (1982) *A Soprano On Her Head*, Utah: Real People Press.
Santayana, George (1896 repr. 1955) *The Sense of Beauty*, New York: Dover Publications.
Sataloff, Robert Thayer (1997) *Professional Voice: The Science and Art of Clinical Care*, (2nd ed) San Diego: Singular Publishing Inc.
Sataloff, Robert Thayer (2005) *Voice Science*, San Diego: Plural Publishing Inc.
Saxon, Keith G, and Schneider, Carol M (1995) *Vocal Exercise Physiology*, San Diego: Singular Publishing Inc.
Schnebly Black, Julia, and Moore, Stephen (1997) *The Rhythm Inside: Connecting Body, Mind and Spirit Through Music*, New York: Sterling Publishing Company Inc. (repr. Alfred Publishing Company 2003). Introduction by Pat Moffitt Cook.
Schumacher, E. F. (1973) *Small is Beautiful: A Study Of Economics As Though People Mattered*, New York: Harper and Row.
Smail, David (1998) *How To Survive Without Psychotherapy*, London: Constable Publishing.
Stevens, Anthony (1990) *On Jung*, London: Routledge.
Storr, Anthony (1996) *Feet of Clay – A Study of Gurus*, London: Harper Collins.
Suzuki-roshi, Shunryu (1973) *Zen Mind, Beginner's Mind*, New York: Weatherhill.
Todd, Mabel (1968, repr. from 1937) *The Thinking Body*, New Jersey: Dance Horizons Inc.
Tolle, Eckhart (2001) *The Power of Now – A guide to spiritual enlightenment*, London: Hodder and Stoughton.
Tomatis, Alfred (1992) *The Conscious Ear*, New York: Station Hill Press.
Tomatis, Alfred (1977) *L'oreille et la vie*, Paris: Éditions Robert Laffont.
Tulloch, Sara (ed.) (1993) *The Readers' Digest Oxford Wordfinder*, Oxford: Clarendon Press.
Verney, Dr Thomas and Kelly, John (1981) *The Secret Life of the Unborn Child*, London: Sphere Books.
Watts, Alan W. (1974 repr. 1985) *The Wisdom of Insecurity – A Message For An Age Of Anxiety*, London: Rider and Company.
Wheatley, Margaret (1999) *Leadership and the New Science*, San Francisco: Barrett-Kochler Publishers Inc.
Zeldin, Theodore (1995) *An Intimate History of Humanity*, London: Minerva.